INTRODUCTION TO
ELECTRICAL INTERFACIAL PHENOMENA

INTRODUCTION TO
ELECTRICAL INTERFACIAL PHENOMENA

EDITED BY

K.S. BIRDI

CRC Press
Taylor & Francis Group
Boca Raton London New York

CRC Press is an imprint of the
Taylor & Francis Group, an **informa** business

CRC Press
Taylor & Francis Group
6000 Broken Sound Parkway NW, Suite 300
Boca Raton, FL 33487-2742

© 2010 by Taylor & Francis Group, LLC
CRC Press is an imprint of Taylor & Francis Group, an Informa business

First issued in paperback 2019

No claim to original U.S. Government works

ISBN-13: 978-0-367-45235-3 (pbk)
ISBN-13: 978-1-4200-5369-2 (hbk)

Visit the Taylor & Francis Web site at
http://www.taylorandfrancis.com

and the CRC Press Web site at
http://www.crcpress.com

Library of Congress Cataloging-in-Publication Data

Introduction to electrical interfacial phenomena / editor, K.S. Birdi.
        p. cm.
    "A CRC title."
    Includes bibliographical references and index.
    ISBN 978-1-4200-5369-2 (hardcover : alk. paper) 1. Interfaces (Physical sciences) 2. Electric double layer. 3. Surface chemistry. 4. Surface energy. 5. Materials--Electric properties. I. Birdi, K. S., 1934- II. Title.

QD555.6.I58I623 2010
541'.37--dc22   2010006625

Contents

Preface

Chemical processes can be characterized in various ways. In one specific case it may concern the presence or absence of electrical charges. This was originally characterized as electrochemistry. Under electrochemistry one finds further subcategories. Electrochemistry is concerned with such simple systems as aqueous solutions of NaCl or more complex systems such as the storage battery. In more complex systems we may add biological cells with charged surfaces. Electrochemistry concerns the charge distribution in the system. The charges near any interface give rise to asymmetric potential. The latter is the subject matter of this book on the electrical interfacial phenomena. The true distance between charges is the main subject of interest.

The text deals with the double layer at the electrode-solution interface. The term *double layer* was used to analyze systems where the charge distribution near interfaces becomes important for the system. Near any interface, there is an asymmetrical charge distribution. This region is where the electrical double layer exists.

The present theory of the double layer depends mainly on dielectric constant concepts, the latter being more typical of 16th- than 20th-century thinking. There are no theories in the double layer that are entirely particulate. The concepts of the double layer deal with the interplay between various layers of electronic charges.

In most physical chemistry textbooks, the subject of electrochemistry is described in its classical fashion. However, due to some major recent developments in surface chemistry, the role of charges at interfaces has become very important (such as batteries, washing processes, adhesion, biological cells (for antibiotics, etc.)). This book presents a picture of the state of an art that is probably on the plateau of further development and stands at a very important stage of application in everyday life.

The subject matter begins with an introduction to the electrical interfacial phenomena, as well as useful examples. Later, more advanced systems are described, and this leads the reader to some comprehensive description of the double layer.

The editor

Professor K. S. Birdi received his undergraduate education in India (B.S., Hons. Chem.) from Delhi University, Delhi, in 1952, and later he traveled to the United States for further studies, majoring in chemistry at the University of California at Berkeley. After graduating in 1957 with a B.Sc., he joined Standard Oil of Richmond, California. Later he joined Lever Brothers in Copenhagen, Denmark, in 1959 as chief chemist in the Development Laboratory. During this period he became interested in surface and colloid chemistry and joined, as assistant professor, the Institute of Physical Chemistry (founded by Professor J. Brønsted), Danish Technical University, Lyngby, Denmark, in 1966. He initially conducted research on surface science aspects (e.g., thermodynamics of surfaces, detergents, micelle formation, adsorption, Langmuir monolayers, biophysics).

During the early exploration and discovery stages of oil and gas in the North Sea, Birdi got involved in Danish Research Science Foundation programs, with other research institutes around Copenhagen, in the oil recovery phenomena and surface science. Later, research grants on the same subject were awarded from the European Union projects. These projects also involved extensive visits to other universities and an exchange of guests from all over the world. Professor Birdi was appointed research professor in 1985 (Nordic Science Foundation), and in 1990 was appointed to the Danish Pharmacy University, Copenhagen, as professor in physical chemistry. Since 1999, Professor Birdi has been actively engaged in consultancy to both industrial and university projects.

Birdi has had continuous involvement with various industrial contract research programs. These projects have been a very important source of information in keeping up with real problems, and have helped in the guidance of research planning at all levels.

Professor Birdi is a consultant to various national and international industries. He is and has been a member of various chemical societies and organizing committees of national and international meetings related to surface science. He has been a member of selection committees for assistant professor and professor, and was an advisory member (1985–1987) of the ACS journal *Langmuir*.

Professor Birdi has been an advisor for about 90 advanced student projects and various Ph.D. projects. He is the author of about 100 papers and articles (and a few hundred citations).

In order to describe these research observations and data, he realized that it was essential to write books on the subject of surface and colloid chemistry. His first book on surface science, along with coauthor D. K. Chattorraj, was published in 1984: *Adsorption and the Gibbs Surface Excess* (New York: Plenum Press). This book remains the only one of its kind. The Gibbs theory is described in this book in extensive detail. The interfacial adsorption of all types of systems are analyzed (soaps and detergents; emulsions; colloidal systems; biological cells: ion-transport and channels; antibiotics). Further publications include *Lipid and Biopolymer Monolayers at Liquid Interfaces* (Plenum Press, New York, 1989), *Fractals in Chemistry, Geochemistry, and Biophysics* (Plenum Press, New York, 1994), *Handbook of Surface and Colloid Chemistry,* (Editor; 1st ed., 1997; 2nd ed., 2003; CD Rom, 1999; 3rd ed., 2008; CRC Press, Boca Raton, Florida), *Self-Assembly Monolayer* (Plenum Press, New York, 1999), *Scanning Probe Microscopes* (CRC Press, Boca Raton, Florida, 2002), and *Surface and Colloid Chemistry: Principles and Applications* (CRC Press, Boca Raton, Florida, 2009). Surface and colloid chemistry has remained his major research interest.

Contributors

Viktor Sergeevich Gevod
Department of Electrochemistry
Ukrainian State Chemical-
Technology University
Dneopetrovsk, Ukraine

Sergey Viktorovich Gevod
Department of Electrochemistry
Ukrainian State Chemical-
Technology University
Dnepetrovsk, Ukraine

Iryna Leonidovna Reshetnyak
Department of Electrochemistry
Ukrainian State Chemical-
Technology University
Dneopetrovsk, Ukraine

chapter one

Introduction to electrical interfacial phenomena

K. S. Birdi

Contents

Introduction

The subject of *chemistry* is taught in high schools and universities. *Electrochemistry* is characterized as a part of general chemistry that relates to charged ions or macromolecules or particles or solids or liquid drops. Charged ions are also found in various other sciences such as physics and biochemistry and geochemistry. However, one finds that electrical charges behave differently when these are situated at or near interfaces. Therefore the state of charges at *interface* needs some special analyses. In the present book the *electrical interfacial phenomena* will be described using the classical electrochemistry and surface chemistry.

In most simple cases such as aqueous solution of a salt, such as NaCl, the Na^+ and Cl^- ions are analyzed by different thermodynamic theories. The addition of NaCl to water (which is neutral) imparts positive (Na) and negative (Cl) ions. The addition of NaCl to water gives rise to various changes in the physicochemical (colligative) properties of the solution (Appendix). This may be:

conductivity
depression of the freezing point
increase of the boiling point

In such solution, the number of positive ions is always equal to the negative ions (as required by the electro-neutrality criteria).

POSITIVE ION	**NEGATIVE** ION
NEGATIVE ION	**POSITIVE** ION

POSITIVE ION   **NEGATIVE** ION
NEGATIVE ION   **POSITIVE** ION
POSITIVE ION   **NEGATIVE** ION

The distance between ions decreases with increasing concentration, as described by the Debye-Huckel (D-H) theory. However, it is found that in small regions the distribution of ions is not equally dispersed. This remarkable observation has given rise to many important consequences, in the simple solution and also in other more complex systems. The distribution of the positive (Na) and negative (Cl) ions throughout the system is found to change with concentration. This means that the distance between a positive and a negative ion decreases with increasing concentration of NaCl (Figure 1.1). Later this will be analyzed and the consequences found to be of much importance in such systems.

It will also be shown that in the case of such solutes as soaps or detergents (surface-active substances or amphiphiles) there is a difference in the adsorption of ions (e.g., positive or negative) at the surface (or interface). The difference in the degree of adsorption would thus give rise to a quantity called surface potential. The monolayers of such surface active substances have been found to be very useful model systems for more complicated systems. In literature one finds a much detailed description related to amphiphiles, since these play a very important role in everyday life (Chattoraj & Birdi, 1984; Adamson, 1999; Birdi, 2003, 2008, 2009; Girault, 2004; Somasundaran, 2006). Another area of interest is where an electrode (metal) is placed in an aqueous solution (Kortum, 1965) (Figure 1.2).

In this case there may be both exchange and adsorption of ions on the electrode. This kind of situation is most important in battery technology. The surface adsorption in such systems has been analyzed by using the Gibbs adsorption theory (Adamson, 1999; Chattoraj & Birdi, 1984; Birdi, 2009). It is found that the concentration of some substances, such as soaps and detergents, is much *higher* at the surface than inside the bulk of solution (as described by the Gibbs adsorption theory). These substances are also called

surface-active agents (substances)
amphiphiles
detergents

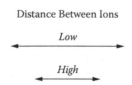

Figure 1.1 Change in distance between Na and Cl ions in an NaCl solution (low concentration; high concentration) (schematic).

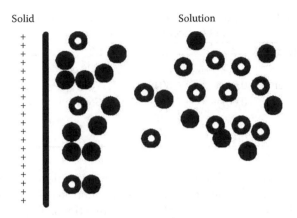

Figure 1.2 Distribution of ions near an electrode with positive charge.

surfactants
emulgators
soaps

Gibbs described the *real system* (which consists of two bulk phases) with an *interphase* between them, by an equivalent system in which the properties of the adjoining phases remain constant up to an interfacial phase, the interface. This is of interest in the case of battery technology. In fact, under the present energy and pollution concern, this technology is expected to expand considerably. Especially, battery technology is related to the application of wind energy and solar energy. Battery can store energy and thus can provide sustained usefulness. Currently one finds a wide range of area where batteries are used (from cars to mobile phones). In fact, battery technology is already well established technology for the CO_2 control and reduction needs in the coming century. In all these systems, the main concern is about how the positive and negative ions are distributed throughout the medium. If one considers an NaCl solution in water, then it is accepted that each positive ion is surrounded by a negative ion (and vice versa) thus giving electroneutrality. The distance between these two different charges will of course depend on the concentration. Based on theoretical derivations, the distance between charges was analyzed by D-H theory (Appendix). The distance (also called the Debye distance), $1/\kappa$, was found to be given as:

$$1/\kappa = 3 \ / \ (\text{concentration of NaCl})^{\frac{1}{2}} \ (10^{-8} \ \text{cm})$$

The values of $1/\kappa$ are found to be dependent on concentration and the charge on the ions (Table 1.1).

Table 1.1 Magnitudes of Ionic Atmosphere, $1/\kappa$, as a Function of Salt Concentration and Type

Electrolyte		1:1	1:2	2:2	1:3
			$(1/\kappa \, (10^{-8} \, cm))$		
Moles/liter	0.0001	304	176	152	124
	0.001	96	56	48	39
	0.01	30	18	15	12
	0.1	9.6	6	5	4

It is thus observed that the *ionic atmosphere* is many times greater in dilute electrolyte solutions than in concentrated systems. However, the differences become lower at higher concentrations. Experiments show that the asymmetry of ions at or near interfaces are of importance in order to understand these systems.

In physics, one describes a capacitor that has two plates of charge separated by some distance. The potential drops linearly from the plate to the other side. However, if a charged metal electrode is placed in a solution, this will not be as simple as a capacitor. This arises from the fact that charges in aqueous media are able to move about and are hydrated ions.

When a solid surface (for example a metal) comes into contact with a solution containing electrolyte, different ionic reactions are found to take place. The solid surface becomes charged due to the difference of ion affinities between the solid and the solution, or the ionization of surface groups. This change in ionic characteristic gives rise to rearrangement of the surrounding ions in the solution. In general, one describes this state of ions by using Gouy–Chapman–Stern (GCS) theory (Appendix). The GCS model gives rise to two layers of specific ions near the surface.

SOLID SURFACE   **STERN LAYER** *DIFFUSE LAYER*

The Stern layer is designated the region next to the charged solid surface (Figure 1.3).

Ions are supposed to be bound due to spacially-adsorbing and Coulombic interactions. The electrical double layer (**EDL**) is the region next to the Stern layer. Ions in the **EDL** region can move freely about in any direction. **EDL** is only present under such situations, and different examples will be considered in this book (Lyklema, 1995; Hunter, 2001; Birdi, 2009) (Figure 1.4).

Experiments show that indeed **EDL** is present at or near any interface.

The significance of the quantify $1/\kappa$ has been found to be important in all kinds of systems:

NaCl solution properties (conductivity, freezing point, etc.)
charged particles (colloids, emulsions, suspensions, etc.)

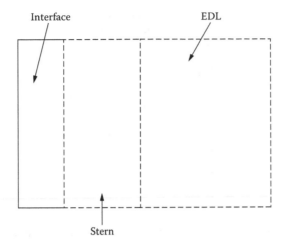

Figure 1.3 Distribution of ions near a charged interface (Stern; Gouy-Chapman) (schematic).

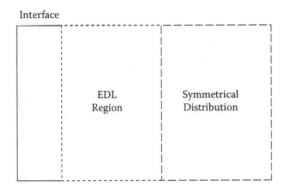

Figure 1.4 Interfacial charge distribution (electrical double layer (**EDL**)) (schematic).

cement industry
paper and ink industry
storage battery (all kinds)
charged macromolecules (proteins, polymers)
biological cells

For example, billions of batteries are commercially produced for various uses and applications (telephones, toys, appliances, cars, radios, instruments, microelectronics). In the case of a battery, one has a positive and a negative electrode as placed in an electrolyte (consisting of fluid phase or gel). The electrode with the positive charge, Figure 1.5, and the state of ions in proximity is shown. It is obvious that near the positive electrode

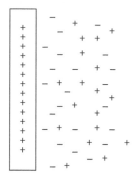

Distribution of Ions Near
Electrode Interface

Figure 1.5 Distribution of ions near an electrode surface with a positive charge (schematic).

there will be a large number of negative ions. The situation near the negative electrode will be just the opposite. It is also obvious that the stronger the potential on the battery, the more counter-ions will be attracted. The asymmetry of distribution thus gives rise to many characteristics, which makes such interfaces unique. Analysis shows, however, that the *number of counter-ions* varies with distance from the electrode.

Thus the state of ions suddenly becomes different from when no electrode was present. The interfacial region where the ions are varying and dependent on the potential of the electrode is the **EDL**. It is thus found that the potential characteristics will depend on the variation of the charges near the electrode. In general, this will also be the case for any other system with a charged surface and its surrounding. The adsorption of ions constitutes the underlying phenomenology of the double layer and contributes to the most experimentally-consistent molecular model of the interface.

The *charge–charge* interactions are found in many everyday systems:

electrolyte solutions
colloidal suspensions
cement industry
paper and ink industry
storage battery
emulsions (oil–water)
biological cells (virus, etc.)
wastewater treatment and pollution control

The two electrodes are chosen such that each has a different chemical potential. The surface charges thus dictate the battery potential. The current flows through the connecting wire from the more negative elec-trode to the more positive electrode (as in the case of a downhill move-ment!). The electricity produced is balanced by ions transported through the electrolyte inside the cell (Appendix). In lead batteries one used flu-ids such as strong H_2SO_4. Nowadays one uses gels so as to stabilize the system on impacts under accidents. In the case of rechargeable batteries the reactions inside the cell are reversed when an opposite potential is applied. Most rechargeable batteries can be cycled more than 1000 times. Battery electrochemistry characteristics are mainly based on the principles of **EDL** on the electrode (Appendix). In general, ions can also be adsorbed on an electrode surface. This can take place under different conditions. The ions may be *specifically adsorbed*. In addition to the latter, an excess of oppositely charged ions are also found to be attracted close to the charged electrode surface. However, due to solvation of ions, these ions cannot approach the surface as closely as the specifically adsorbed ions. These ions are found to be distributed in a diffuse layer. It is thus found that near the surface of an electrode (Figure 1.5) one has the fol-lowing ions:

electrode surface with charge
diffuse layer of opposite charges

This state of distribution of ions, in general, is called the **EDL**. Accordingly, one finds that **EDL** has been analyzed extensively in the literature.

This book deals with those interfacial phenomena that are related to charges in the liquid phase or the solid phase. As will be shown, the dis-tribution of charges in any given phase (liquid or solid) is not symmetrical around or near a charge. This situation becomes even more asymmetrical near any **interface (air–liquid, liquid–liquid, solid–liquid).**

The role of electrical interfacial phenomena in various industrial pro-cesses and products is well established. Some typical and important areas are as follows:

paint industry (electro-deposition)
aircraft industry
auto industry
agrochemical
photographic
printing ink
detergent and washing
dyestuffs
ceramic
cement

biology
emulsion technology
energy industry
oil production
coal technology
pollution control
wastewater treatment

Typical examples of such systems will be described.

Colloidal systems are found in different areas of everyday life. In the wastewater treatment the colloidal suspension is destabilized through the application of **EDL** science. The Derjaguin-Landau-Verwey-Overbeck (**DLVO**) theory is mainly used in such systems treatment. Since drinking water is becoming a very critical aspect for the survival of mankind, it is obvious that this area of science will need much research in the future. Another area where colloidal particles are involved is the paint and paper industry. In these systems also the understanding of such suspensions is evaluated by applying various surface and colloid science theories. The development of nano-scale particles (nano-technology) is another new application of **EDL** that is of recent origin.

Emulsion technology is one of the most important areas where the *interfacial charges* play an important role. Oil and water do not mix, as is well known. However, if the high interfacial tension (ca. 50 mN/m) at the oil–water phase is considerably reduced (less than 0.1 mN/m) by addition of suitable emulsifiers and so forth, then the system becomes stable for a longer time. Many of these emulsifiers used to stabilize emulsions are ionized so the emulsion droplets exhibit an electric charge. In these oil–water emulsion systems, the presence of such charges at the interface will lead to the formation of an **EDL**.

Emulsion

OIL SURFACE (CHARGE). WATER SURFACE (CHARGE).

The nature of **EDL** will determine the stability characteristics of the given emulsion. The stability of the system would then be dependent on:

EDL repulsion
van der Waals attraction

When two charged emulsion droplets approach each other, there is a repulsive interaction when the diffuse layers begin to overlap (as depicted below). The magnitude of this repulsion energy increases as the region of the overlap increases and the kinetic movement of the droplets enhances the movement.

OIL DROP   INTERFACIAL CHARGE REGION
INTERFACIAL CHARGE REGION—INTERFACIAL CHARGE REGION =
EDL REGION
OIL DROP (1)   **EDL** REGION   **OIL DROP** (2)

It was found that the **EDL** region is dependent on various factors (Chapter 2).

For example, the stability of two charged particles (such as in wastewater treatment), is dependent on the following:

concentration of electrolytes in the surrounding solution
the charge valance, Z_{charge}, of the counter-ion (actually the stability is related to $Z_{charge}{}^{6}$)

This shows that the interfacial charge region is not as simple as one may imagine as a first approximation. Electrostatic repulsion is probably a more common mechanism for the stabilization of emulsions and *van der Waals* (**vdw**) forces than any other force. Soaps, detergents, and many emulsifiers operate in this fashion to stabilize aqueous dispersions and emulsions. For example, many non-aqueous dispersants stabilize dispersions by electrostatic charges (engine oils).

Another area of interest is the **biological systems**. In general, all biological cells are negatively charged. The structure of a cell is basically as follows (Figure 1.6):

outer layer—lipid–bilayer membrane (**BLM**)
BLM contains proteins (*membrane proteins*)
inside the cell—the composition of the fluid inside the cell is not the same as the fluid outside

Both the lipids (lecithins) and the membrane proteins may exhibit charges. This leads to the same kind of interfacial considerations as

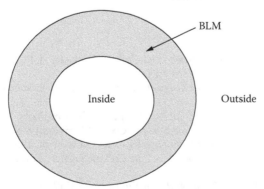

Figure 1.6 Biological cell (**BLM** and the proteins) (schematic).

mentioned above for other systems. The cell–cell repulsion will be thus dependent on the surface charges (i.e., surface potential). These potentials can be measured by *electrophoresis*.

BLM exhibits very characteristic properties. One of these is where the composition of the fluid inside the cell is different from that of the outside.

INSIDE FLUID   **BLM**   OUTSIDE FLUID

However, *antibiotics* are used to destabilize this, which leads to the destruction of the cells. For instance, valinomycin (antibiotic) is able to attach to the **BLM** and creates a *channel* for the free transport of K^+ ions only (see Chapter 3).

As an example, the bee venom *melittin* is a very unique membrane protein (26-amino acid chain). It can penetrate into the **BLM** very easily because it is highly hydrophobic (very low solubility in water). However, the amino acid composition is unique, which induces a positive charge very near or at the cell interface. The proximity of this specific charge disrupts the **BLM**. The model membrane experiments have shown how these membrane proteins behave in real systems (see Chapter 3). Experiments have shown that the most useful procedure is to use the *monolayer* method, which was used by Langmuir (who was awarded the Nobel Prize for using this model membrane method) (Birdi, 2009). Melittin has recently been found to exhibit therapeutic activity. It has been reported to show powerful anti-inflammatory properties (for example, against rheumatoid arthritis).

In general, one finds a whole range of membrane proteins exist (which in general consist of about 20 to 25 amino acids). These specific proteins have been extensively investigated in the current literature. These membrane proteins are characterized as *ion carriers*. This arises from the fact that the lipid phase of the **BLM** inhibits all ion transport in a cell. The model monolayer studies have added much useful information about the cell function and mechanisms.

References

Birdi, K.S., ed., *Handbook of Surface and Colloid Chemistry*, CRC Press, Boca Raton, FL, 2003 (2nd ed.); 2008 (3rd ed.).

Birdi, K.S., *Surface Colloid Chemistry, Principles and Applications*, CRC Press, Boca Raton, 2009.

Chattoraj, D.K. and Birdi, K.S., *Gibbs Adsorption*, Plenum Press, New York, 1984,

Hunter, R.J., *Foundations of Colloid Science*, Oxford University, New York, 2001.

Kortum, G., *Treatise on Electrochemistry*, Elsevier, New York, 1965.

Lyklema, J., *Fundamentals of Interface and Colloid Science*, Vol. II, Academic Press, London, 1995.

Somasundaran, P., ed., *Encyclopedia of Surface and Colloid Science*, 2nd ed., CRC Press, Boca Raton, FL, 2006.

chapter two

Interfacial charge and basic electrical double layer (EDL)

K. S. Birdi

Contents

Introduction

Charged chemical species are found in everyday life. *Electrochemistry* deals with charges in aqueous media. All the systems which will be discussed in this book are related to aqueous media. This is due to the fact that charged ions, such as Na^+ or Cl^-, or soaps or macro-ions, can only exist in aqueous media. Of course, one finds systems where charges are present even in the absence of water (such as static-charged systems, or thunderstorms), but these systems are out of the scope of this book.

In aqueous media, the charges are able to move about, thus these systems are completely different from the static systems. Therefore, most of the systems will be related to changes observed in aqueous media as a function of ions (concentration, type, size). The main theme in this book is related to the spatial distribution of charges near the *interface*. This will be shown as being a very special case regarding the spatial distribution of ions at and near the interface (such as when a metal electrode is placed in an electrolyte media, such as in a storage battery). The electrical phenomena at interfaces will be therefore related to the *distribution of charges* (Figure 2.1).

The understanding of the charges at interfaces is of importance in many everyday phenomena:

electrochemistry (electrokinetics)
microfluids
colloidal systems (emulsions, pollution)
solid–liquid interfaces (battery, electrometers)

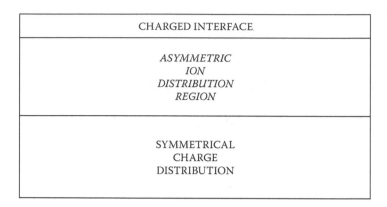

Figure 2.1 Asymmetric distribution of charges near an interface (air–liquid, liquid–liquid, solid–liquid).

The latter is called electrical double layer (**EDL**) which is found to be of importance in many different systems as found in everyday life. The Stern model defines the charges as those that are near a charged surface and are strongly fixed, whereas away from the Stern plane the ions are asymmetric and are defined as the **EDL**.

This may be depicted most simply as follows. At the interphase of two different phases with charges (Figure 2.2), there will be overlap of potential.

Phase I   **POTENTIAL OVERLAP** Phase II

In this case there are two aspects that are important. First, the *asymmetric* potential at the interface will depend on the extent of the charge and the surrounding electrolyte concentration. Second, the overlap between two charged particles will depend on the distance between the particles and the latter property. For example, if in a particular case a suspension of charged particles is stable, then it can be made unstable if one can reduce the overlap distance (by adding more electrolyte). The system will be stable or unstable, depending on the character of the potential overlap, also called **EDL**.

It is important to describe the state of pure liquids and solids before undertaking the analyses of other complex systems. This helps one to appreciate the molecular picture of such systems, especially regarding the state of molecules at any interface.

The basic physicochemical description of liquids and solids is generally taught in all high school– and university-level courses. However, the state of the *molecules at the surfaces* is generally not mentioned in the physics or chemistry textbooks as extensively as one would desire. Let us therefore consider a simple approach about the state of molecules in

Charged
Particle

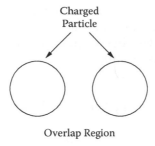

Overlap Region

Figure 2.2 Overlap of the **EDL** of two charged particles.

the bulk of a liquid as compared with the surface molecules. If we take a simple liquid such as water as an example, then we have the following situation. A beaker (a closed container) with water is found to be at equilibrium with its surroundings, such that the number of molecules evaporating per second is the same as those that are condensing. That means there is a very active (dynamic) movement of molecules at the surface (Adamson, 1999; Birdi (a), 1989, 1999, 2002; Birdi (b) 2003, 2008, 2009). Further, inside the bulk phase a molecule of water is symmetrically interacting with its neighbors and thus is kept in a more or less *fixed* situation. The molecules in liquids move just about 10% more than in solids. The distance between molecules in the liquid state is much lower (approximately one-tenth) than in the gas phase. This can be estimated by the following example.

Water data (25°C)
Volume of *gas* per mole = 22.4 liters/mole
Volume of liquid per mole = 18 cc/mole
Ratio of volume gas:liquid = 22400:18 = ca. 1000

One thus finds that in the gas phase the distance between gas molecules is roughly 10 ($\sim 1000^{1/3}$) times larger than in the liquid phase. Since the dividing line between these two different phases is very sharp (as seen by eye!), one can thus expect that there must be a large molecular rearrangement. It is of interest to mention that the volume of a solid is generally 10% less than that of a liquid. Thus the main difference between a solid and a liquid is mainly that the molecules in the former are almost fixed, while in the latter there is some movement. The situation at such interfaces as liquid–gas thus becomes important if solutes (such as ions) are added that may adsorb at this interface. The outermost layer of an aqueous solution surface is conventionally assumed as being devoid of ions. This is argued due to the fact that most salts in water give an increase in surface tension. Recent studies have however indicated that some large ions (such as I$^-$) exhibit asymmetrical charge distribution (Petersen and Saykally, 2005). The distribution of ionic substances at such interfaces has been shown to play an important role. This may be depicted as follows (first few layers at the interface).

If one has a system consisting of:

water
NaCl (which dissociates into Na+ (+) and Cl– ions (–))

then one can describe this system as follows:

Pure water (depicted as **w**):

wwwwwwwwwwwwwwwwwwwwww
wwwwwwwwwwwwwwwwwwwwww
wwwwwwwwwwwwwwwwwwwwww
wwwwwwwwwwwwwwwwwwwwww

Water (w) + NaCl (using + for Na, and – for Cl):

www–w+w–w+w+w–w–w+w–w+wwww–www+
www–www+www–www+www–www+
www–www+www–www+www–www+
www–www+www–www+www–www+

In NaCl solution there will be equal number of + and – ions. This is required by the electrical neutrality criteria. The situation is not symmetrical near or at the interface. Measurements have shown that at the surface of NaCl solution there can be more + ions (Na) than – (Cl) ions:

Surface layer:
w+ww–www+ww+w+wwww+ww+w–www+

This means there will be a *local positive* potential, as compared to the whole system with neutral potential. Analogous to this, one has found that in KCl solutions, the interface is almost neutral. This has been used in systems with KCl bridges in electrochemistry. Studies have shown that the surface iodide concentration in HI is larger than in the case of NaI or KI. The iodide concentration of HI was 55 % larger than in NaI. The latter observation thus has been found to have important consequences in systems where surface charges (surface potential) will be dependent on the electrolytes. In other words, even though most properties of two different electrolytes might be the same in the bulk phase, at the interface one will always find differences.

Let us consider another system consisting of:

water
sodium dodecyl sulfate (**SDS**)

Surfactants have found widespread applications in various industrial systems. Commercially available surfactants are found in a range of products including motor oils, lubricants, pharmaceutical, detergents and soaps. In addition these molecules are used in the manufacture of many common materials such as plastics and textiles and are also used in the oil recovery processes. One also finds many biological systems where surfactants are used at different levels. Besides such vast range of application area, the structure and their exact function at molecular level at interfaces (such as: oil–water; air–water; and solid–water) are not completely understood.

It is useful to consider systems consisting of water solutions of detergents, as a general example. It is known that SDS, being a detergent (anionic), reduces the surface tension of water even in very minute concentrations. From such experiments it is known that the distribution of SDS, which dissociates into SD^- and S^+ ions (SDS dissociates in the same manner as NaCl), is asymmetric at any interface. This system will be something like NaCl but with a big difference around the interface, as depicted below (first few layers of interface).

water...depicted as **w**
SD–...depicted as **D**
S...depicted as **+**

wwDwDwDww+*wDwwDwDwwDwDwwDwww*+
wwDwDwww+wDwwDwww+wwwDwDww+Dw
wwwDwww+wwwDwww+wwwDwww+wDwwD
wwwDwww+wwDwDwDww+wDwwDwwDw

This shows that the surface of such a system changes in charge from almost neutral to *negative*, because the number of SD^- ions has always been found to be greater than S^+ (as found from experiments). This has been determined from direct surface potential measurements. The critical concentration at which the surface is saturated with SD^- ions is found to take place at 0.008 mole/liter (in water at 25°C). Analogously, if one analyzes a cationic detergent system, such as cetyl trimethyl ammonium bromide (**CTAB**), one finds that the surface potential changes to *positive*. **CTAB** dissociates in water to: CTA^+ and Br^- ions. In literature one finds extensive reports on such studies (Birdi, 2009).

Thus one finds that the neutral charged surface of a solution under given conditions may change to negative potential (in the case of SDS system) or positive (in the case of **CTAB** system) (Figure 2.3).

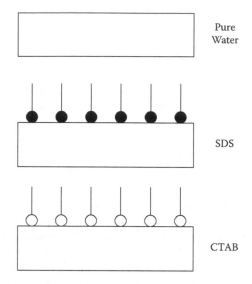

Figure 2.3 States of surface charge (surface potential) of water (pure) neutral charge; (with **SDS**) negative; (with **CTAB**) positive.

It is of interest to note that only SDS can be used for washing (clothes, hair shampoo). On the other hand, CTAB cannot be used for these applications. This difference arises from the charges at the interface in the two systems.

The state of the surface in such systems as SDS–water, one thus finds that following properties of the solution change:

surface tension (γ) decreases with SDS concentration (C_{SDS}).
SD– ions preferentially adsorb at the surface.

From Gibbs adsorption theory (Chattoraj & Birdi, 1984; Birdi, 2009), one finds that *change in surface tension is proportional to adsorption of surface active substance (such as SDS).*

The Gibbs adsorption equation (Appendix) is written as:

$$-d\,\gamma\,/\,d\,(Log\,(C_{SDS}) = R\,T\,(\Gamma_{SDS})\qquad(2.1)$$

where γ is surface tension (mN/m), Γ_{SDS} is area/molecule of SDS molecule at surface. One can estimate the magnitude of Γ_{SDS} from the plots of γ versus Log C_{SDS}. In general, the surface tension of water decreased as follows:

Pure water   72 mN/m   (25°C)
SDS concentration 2.3 g/liter   25 mN/m

The magnitude of Γ_{SDS} was found to be in the range of 40 to 60 A^2/molecule. This agrees with data from other methods.

In systems with charged species, Γ is thus the area per charge. This quantity allows one to investigate these systems with respect of the **EDL** (see Chapters 2 and 4).

Under this condition, one finds that the interface changes charge from *neutral* to *positive* or *negative*. But one also notices that the interface is not neutral, and the asymmetry near the interface is the region of interest in this book.

Analogous to this, one also finds that if a solid is immersed into a liquid, then the following may exist:

    solid is *neutral* in charge
    solid is *negatively* charged
    solid is *positively* charged

This is most important in technologies such as the making of batteries and the like. In other systems, such as colloidal systems, the situation of surface charges is even more complex (Appendix B).

For instance, if one considers a suspension of silica particles in water, situations such as the following may exist:

    Silica is negatively charged.
    On addition of a cationic detergent, such as CTAB, the surface charge of
       silica changes to zero and subsequently to positive.

This shows that one can obtain a variety of charged silica varying from:

SILICA PARTICLE CHARGE:
negative / zero / positive,

depending on the concentration of the added cationic amphiphile (Brown, 1999). This structure is termed **EDL**. The pH effect showed that the magnitude of surface charge was as follows:

    positive charge   pH below 5
    zero charge   pH = 5
    negative charge   pH above 5

A similar situation is observed in other macromolecule systems, such as a protein and charged amphiphile.

At any interface one will thus expect an electric potential difference, where any two electrically conducting phases are in contact (Figure 2.4).

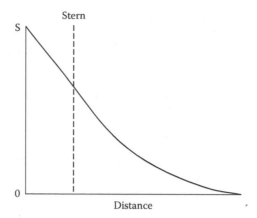

Figure 2.4 Electrical double layer (**EDL**) at a phase boundary.

This potential change has been the subject of extensive studies in the literature. Let us consider a system as follows:

metal electrode in contact with an electrolyte solution

The metal surface exhibits a (for example, positive charge) potential, ψ_o, which reduces to zero at some distance away from its surface. The characteristics of such an electrode are found to be dependent on the following parameters:

magnitude of surface charge
magnitude of distance from surface where the value of surface potential
    becomes zero.

The surface charge (positive) will decrease as one moves away from the surface. According to the most simple model, as suggested by Stern, one suggests that at some distance, Figure 2.4, the surface potential will decrease to zero (Birdi, 2009). However, experiments have shown that this decrease is different for different systems because it was found to be dependent on the characteristics of the charges present in this region. From experimental data this was found to be unrealistic, mainly because in aqueous systems one has to consider the presence of water molecules as associated with all ions. This would therefore suggest that the surface potential will decrease in a more realistic manner as shown in Figure 2.5.

The number of opposite charges (negative–anioins) will be *greater* near the metal surface, while there will be *lesser* positive (cations) ions in the aqueous media. The *thickness* of this asymmetry region will be dependent on the Debye–Hückel length, $1/\kappa$. It has been found that the magnitude of surface potential, ψ, varies with distance, x, as follows:

$$\psi = \psi_o \exp(-\kappa x) \tag{2.2}$$

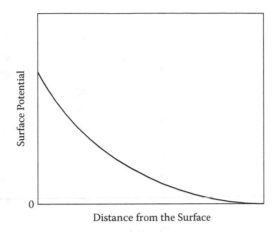

Figure 2.5 The Gouy–Chapman model of the surface potential.

Thus the **EDL** region is present only near the interface (e.g., air–liquid, liquid–liquid, solid–liquid). The **EDL** is absent as one moves away from the surface where the number of positive ions is the same (and symmetrically distributed) as the negative ions. The **EDL** region thus decreases as the value of $1/\kappa$ decreases with increasing electrolyte concentration. The variation of surface potential is thus related to κ as given by the above equation.

It is common experience that if one blows air bubbles in pure water, no foam is formed. On the other hand, if any surface active substance (such as soap, detergent, protein, etc.) is present (even in minute quantity) in the system, adsorbed surface-active substance molecules at the interface give rise to foam or soap bubbles.

One of the most convincing examples one finds is the case of soap bubbles and electrolyte concentration.

```
(Soap bubble film structure)
SOAP   WATER PHASE SOAP
SOAP   WATER PHASE SOAP
SOAP   WATER PHASE SOAP
SOAP   WATER PHASE SOAP
```

The *thickness* of the soap film is sum of soap molecule (nm) + water phase (µm) + soap (nm). This varies from micrometer to nm (Birdi, 2009). A soap bubble is a structure of a double layer of soap molecules with water in between (Figure 2.6).

It has been found from experiments that the thickness of soap bubbles decreases as one adds electrolyte to the system. This arises from the

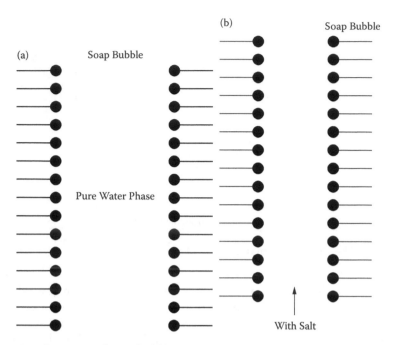

Figure 2.6 Structure of soap bubble without (a) or with (b) added electrolyte.

fact that the magnitude of **EDL** decreases such that the charge–charge repulsion between the layers of soap decreases with added electrolyte. The surface potential on each soap layer will repel other soap layers. The overlap region will be larger in low-electrolyte solution. On the other hand, the overlap region will be compressed in high-electrolyte solution. Experiments have shown that the equilibrium thickness of a soap film, t_{film}, is related to $1/\kappa$ as follows (Birdi, 2009):

$$t_{film} = 6 / \kappa + 2 \text{ (length of the soap molecule)} \qquad (2.3)$$

This shows that in a very simple system such as soap bubble film, the **EDL** behavior is important in prediction of its structure and stability. Foam technology is important in such systems as washing, firefighting, emulsions, and others. In sodium oleate solution, the following was found:

$$t_{film,oleate} = 600 + 2 (30) = 660 \text{ Å}$$

where $6 / \kappa = 600$ Å, in a 0.001 M solution. This value agrees reasonably with the measured value of 700 Å (Birdi, 1989).

Another important application of foam technology has been the wastewater purification (Birdi, 2009). The main principle is to be able to collect the monolayer of any surface-active substance, which is preferentially adsorbed at the interface. The SAS is indicated as **I** and water is indicated as **W**:

INTERFACE
IIIIIIIIIIIIIIIIIIIIIIIIIIIIIIIIIIIIIIIIIIIII
WWWWWWWWWWWWWWWWWW
WWWWWWWWWWWWWWWWWW

Thus, if one could remove the IIII layer (by suction at the surface), then the concentration of the SAS will decrease. Since surface will get saturated with a new lipid layer after suction, one may repeat this process many cycles over. This is actually used in bubble-foam purification processes (Birdi, 2009). Further, one can notice that even very small concentrations (mg/liter or ppm) can be removed with this foam method, provided the contaminant is a SAS.

This can be shown as follows:

IMPURE SYSTEM WITH SAS (as I)
IIIIIIIIIIIIIIIIIIIIIIIIIIIIIIIIIIIIIIIIIIIIIIII
WWWWWWWWWWWWWWWWWW
WWWWWWWWWWWWWWWWWW

AFTER SUCTION
WWWWWWWIWWWWWWWWW
WWWWWWWWWWWWWWWWWW

AFTER NEW EQUILIBRIUM
IIIIIIIIIIIIIIIIIIIIIIIIIIIIIIIIIIIIIIIIIIIII
WWWWWWWWWWWWWWWWWW
WWWWWWWWWWWWWWWWWW

One can repeat this process, thus achieving the removal of any SAS with very low concentration (as low as parts per million (ppm)). This is obviously a very useful method, since it requires chemical application and the purification is a very fast process. Bubble-foam purification is in fact based on the same principle. As bubbles form at the surface, the SAS is mainly retained in the thin-liquid film. The bubbles can be removed and the process can be carried out until most of SAS is removed (in some cases over 99% contaminant can be removed).

There are some well-known systems where the interfacial electrical charges are of common knowledge. In the following a few examples are given to explain these ionic distributions.

Surface charge determination of glass and silica surfaces

In many industrial applications the surface charge, for example, at the surface of glass or silica, is of much interest. Glass is one of the most commonly used materials in everyday life. The properties of glass are related to its surface characteristics (among others, surface charge) (Behrens and Grier, 2001). In general, when silica and silicate glass surfaces are immersed in water, it has been found that these surfaces acquire a negative charge. This depends on the charges at the surface and the counter-ions in the bulk solution. The glass surface is reported to acquire approximately $-2000\ \varepsilon/\mu m^2$, where ε is the elementary charge. The dissociation on silica surfaces is as follows:

$$SiOH = SiO^- + H^+ \tag{2.4}$$

The effective charge density, ψ_{eff}, was found to fit the following equation:

$$\psi_{eff} = e\ \kappa\ /4\ \pi\ (\psi_{eff})\ (1 + 1/\ (\kappa\ \mathbf{a})) \tag{2.5}$$

where ψ_{eff} is the surface potential, $\mathbf{a}$ is the radius. Similar procedure was found to apply to other charged surfaces.

Electrical interfacial properties of emulsions

In our discussion of an everyday, real example, the *emulsion technology* is one of the most important applications.

As is well known, oil and water do not mix. If one adds oil to water and shakes the system, the oil drops break up, but after a very short time these coalesce again and form a continuous phase. However, if one adds a suitable emulsifier, then the interfacial tension is decreased and this gives rise to very small oil drops. This may give rise to a very stable emulsion system (Figure 2.7).

In some emulsions the oil drops may exhibit charges if the emulsifier is an ionized molecule (such as a soap molecule). The charge may be positive or negative. It is found that in oil–water (O/W) emulsions, these charges lead to **EDL** characteristics near the oil drop surface. The negatively charged oil drops will thus attract opposite charges and repel ions of the same charge. This leads to the formation of a diffuse layer throughout which the asymmetrical distribution of charges exits. At some distance away from this region, there is symmetry around each charge, and hence **EDL** is absent. As mentioned elsewhere, if one increases the concentration of electrolyte in the water phase, then the **EDL** region is

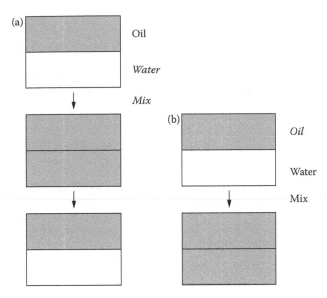

Figure 2.7 Oil–water systems: (a) oil and water; (b) oil–water–emulsifier.

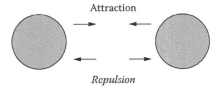

Figure 2.8 Potential energy between two charged oil drops in an emulsion as a function of distance of separation.

decreased (this is due to the decrease in $1/\kappa$). This leads to the observation that two interfaces can come closer than when the magnitude of a double layer is large. This means that every charged interface exhibits a diffuse layer. The shape of this layer is of interest, since this determines the characteristics of the system.

The state of interaction between two charged oil drops is thus described as follows: As two charged drops approach each other, at some distance the diffuse layers begin to interact due to overlap of the potentials. The repulsion energy increases as the overlap energy increases. The attraction forces arising from van der Waals forces (Appendix) will increase at very short distances (much shorter than electrical repulsion region). The kinetic energy present due to the relative motion of the droplets thus determines the total stability of the system. This is depicted in Figure 2.8.

References

Adamson, A., *Physical Chemistry of Surfaces*, 5th ed., Wiley Interscience, New York, 1999.

Behrens, S.H., and Grier, D.G., *Condensed-Matter*, vol. 2, May 8, 2001.

Birdi, K.S., *Lipid and Biopolymer Monolayers at Liquid Interfaces*, Plenum Press, New York, 1989.

Birdi, K.S., *Self-Assembly Monolayer Structures of Lipids and Macromolecules at Interfaces*, Plenum Press, New York, 1999.

Birdi, K.S., *Scanning Probe Microscopes*, CRC Press, Boca Raton, FL, 2002.

Birdi, K.S., *Handbook of Surface and Colloid Chemistry*, CRC Press, Boca Raton, FL, 2003; 2008.

Birdi, K.S., *Surface and Colloid Chemistry*, CRC Press, Boca Raton, FL, 2009.

Brown, G.E., Metal oxide surfaces, *Chem. Rev.*, 99, 77, 1999.

Chattoraj, D.K., and Birdi, K.S., *Gibbs Adsorption*, Plenum Press, New York, 1984.

Petersen, P.P., and Saykally, R.J., Hydronium concentration at the water surface, *J. Phys. Chem.*, 7976, 109, 2005.

chapter three

Electrical aspects of surface pressure in amphiphilic monolayers

V. S. Gevod, I. L. Reshetnyak, and S. V. Gevod

Contents

Introduction

Amphiphilic monolayers (surface films formed at air–water interface) are known as the useful models for study mechanisms of different biophysical and biochemical phenomena in living cells. They can also provide important knowledge regarding properties of thin amphiphilic arrangements (fatty acids, lipids, proteins, and mixed films) in agricultural, pharmaceutical, and food science applications. In order to receive this knowledge one needs to know the algorithm for evaluation of fundamental forces in the space of monolayers. These forces depend upon packing density of monolayer molecules, their three-dimensional structure, concentration of subphase electrolyte, etc. Below describes the different approaches for analysis of the monolayer state.

The main attention will be devoted to evaluation of monolayer surface pressure. In any amphiphilic monolayer the magnitude of surface pressure is equal to the difference between the surface tension of "pure" subphase and surface tension of subphase covered by amphiphilic molecules, that is:

$$\Pi = \gamma_0 - \gamma_1, \tag{3.1}$$

where Π – surface pressure of monolayer, γ_0 – surface tension of "pure" subphase, γ_1 – surface tension of subphase covered by monolayer.

Kinetic component of surface pressure

Kinetic forces arise in any amphiphilic monolayer due to the thermal motion of their molecules or units. The value of kinetic component of surface pressure depends upon packing density of monolayer molecules, their dimensions, and temperature. The numerous investigations have proved that the equation of ideal two-dimensional gas satisfactorily describes the kinetic component of surface pressure in «gaseous» monolayers [1]:

$$\Pi_{kin} = \frac{kT}{S_i}, \tag{3.2}$$

where S_i – specific area in monolayer, k – Boltzman constant, T – absolute temperature.

In extended and condensed monolayers a correction must be made for excluded area S_0. It is equal to a cross-section of monolayer molecules in the plane, parallel to subphase. Thus the equation (3.2) becomes:

$$\Pi_{kin} = \frac{kT}{S_i - S_0}. \tag{3.3}$$

The value S_0 appearing in equation (3.3) one can take from the measurements of specific area of monolayer in collapse point.

Dispersion component

In comparison with the kinetic component, the sign of the dispersion component is negative. Its absolute value is proportional to the sixth degree of the distance between molecules of monolayer and described by expression:

$$\Pi_{disp} = \frac{B}{a^6},\tag{3.4}$$

where B – constant of dispersion interaction and a – distance between molecules.

Under real conditions the dispersion force interaction between monolayer molecules is negligible up to the close packing density [1,2]. Therefore the surface pressure is noticeably affected by Π_{disp} only in condensed monolayers.

Electrostatic component

The Coulomb forces are the powerful reason for surface pressure in ionized monolayers. These forces act between charges inside the monolayer, and between monolayer charges and subphase ions. Coulomb forces are proportional to the second degree of distance between the charges. They have a marked effect on the surface pressure in monolayer, depending upon the structure and ionic concentration of subphase. The excess of Π upon Π_{kin} is registered when the ionized amphiphilic monolayer molecules contact with subphase, which consists of electrolyte solutions of 1–1 type with low ionic strength [3]. The reduction of surface pressure in ionized monolayers in comparison with non-ionized monolayers sometimes takes place at their interactions with some multicharged inorganic anions and cations [4]. To analyze the dependence of the electrostatic component of surface pressure upon the kind of monolayer substance, ionic composition, and subphase concentration, some approaches described in the literature were offered [2,5–7].

Electrostatic component of surface pressure in monolayers with dipole distribution of charges is usually determined with regard to the energies of rotary and oscillatory movements of amphiphilic molecules. This analysis can be carried out by means of a cut-out disk model considering the discrete charges interactions [7].

In the case of ion–ion interactions, the electrostatic component is usually identified with the change of free energy at the electrolyte–air interface, which takes place due to ionization of monolayer molecules [5]. The most widespread approaches in this case are associated with the names of Donnan [2], Davies [8,9], Phillips and Rideal [10,11], Bell, Levin, and Pethica [3,12–18]. Below we shall deal with each of these approaches and we shall estimate borders of their applicability.

Donnan approach for ion–ion interaction analysis in monolayers

This approach consists in a finding of osmotic pressure in the surface layer of electrolyte disposed under amphiphilic monolayer. According to the Donnan theory the surface zone in electrolyte–monolayer–air system, which is situated between the geometrical interface and subphase volume, is represented as a thin hypothetical film with final thickness δ. Thus, according to the theory, amphiphilic molecules are displaced on one of the film borders, and the other border is conventionally placed where the concentration of inorganic anions and cations in subphase become identical with each other. Within the examined film the quantities of surface-inactive ions are different from each other. The structure of their concentration in considered layer is determined by Donnan equilibrium. Donnan effect assumes the ions' replacement from monolayer molecules, where the charge sign of these ions coincides with the charge sign of monolayer. And, in that way, the concentration of opposite-sign ions is increased. Redistribution of inorganic ions in the boundary layer with thickness δ causes generation of Donnan potential (φ_d).

For the system with concentration of surface-active cations (C_Q^+), located on air–water interface, the ratio of surface-inactive ions' concentration on this interface (C_s^+ and C_s^-) and in subphase (C^+ and C^-) is connected with Donnan potential by follow equations:

$$C_s^+ = C^+ \exp\left(-\varphi_d \frac{e_0}{kT}\right) \tag{3.5}$$

$$C_s^- = C^- \exp\left(+\varphi_d \frac{e_0}{kT}\right) \tag{3.6}$$

In the case of symmetric electrolyte the concentrations of C^+ and C^- are equal to volumetric concentration of subphase electrolyte (C).

Equilibrium condition of ions' concentration in subphase bulk and in monolayer is as follows:

$$C^+C^- = C_s^+C_s^-, \tag{3.7}$$

and the condition of electro neutrality preservation of the system assumes that

$$C_s^- = C_s^+ + C_s^+. \tag{3.8}$$

Thus,

$$C_Q^+ = \frac{C^2}{C_S^+} - C_S^+.$$ (3.9)

Combining the expression for C_Q^+ with (3.5), (3.6), and (3.7), we shall get:

$$C_Q^+ = \left[\exp\left(\frac{e_0 \varphi_d}{kT} \right) - \exp\left(-\frac{e_0 \varphi_d}{kT} \right) \right] \cdot C = 2Csh\left(\frac{e_0 \varphi_d}{kT} \right).$$ (3.10)

As far as it is possible to present formally the concentration of surface-active ions in monolayer as the ratio of surface surplus of these ions (C_Q^+) to the depth δ, where the concentration of surface-inactive ions is leveled, the interrelation between the Donnan potential and surface concentration of ionized surface-active substances is described by the equation [2]:

$$\varphi_d = \frac{kT}{e_0} Arsh \frac{1}{2N\delta S_i C},$$ (3.11)

where S_i – the area accounted for separate surface-active ion in monolayer, N – Avogadro's number.

So, it is possible to calculate the Donnan contribution to the surface pressure of ionized monolayer on the basis of (3.11). But inasmuch as the correct calculation is too complicated, the limiting relation of osmotic theory is usually used [2]:

$$\Pi_{osm} V = RTN_2^s$$ (3.12)

or

$$\Pi_{osm} = RT\frac{n_2^s}{\Omega},$$ (3.13)

where Π_{osm} – osmotic pressure, V – molar volume, R – universal gaseous constant, N_2^s – molar fraction of all ions at the interface, n_2^s – the surplus of all ions in surface layer, Ω – the monolayer area.

And since the excessive concentration of all ions in the layer with thickness δ submits by equation:

$$C_{surpl} = C_Q^+ + 2C\left[ch\frac{e_0 \varphi_d}{kT} - 1 \right],$$ (3.14)

the general solution of (3.13) and (3.14) gives the expression for surface pressure:

$$\Pi = RTT_Q^+ + 2C\delta RT\left[ch\frac{e_0\varphi_d}{kT} - 1\right].\tag{3.15}$$

The second term in (3.15) represents the osmotic contribution, and the first one corresponds to kinetic component. If the thickness δ is equal to Debye shielding length (χ), where concentrations of cations and anions in subphase become identical, then (3.15) with regard to (3.11) will get the form:

$$\Pi = RTT_Q^+ + \frac{kT}{e_0}A\sqrt{C}\left[chArsh\frac{2e_0}{S_1A\sqrt{C}} - 1\right],\tag{3.16}$$

where

$$\chi = \sqrt{\frac{2CNe_0^2}{\varepsilon_0\varepsilon_1 kT}},\tag{3.17}$$

$$A = \sqrt{2NkT\varepsilon_0\varepsilon_1}.\tag{3.18}$$

And when the ratio $2e_0/Si\sqrt{8CNkT\varepsilon_0\varepsilon_1}$ is so great that it is possible to consider that

$$chArshX = \sqrt{1 + X^2} \cong X,\tag{3.19}$$

then one will get

$$n = RT\partial_Q^+ + \frac{kT}{Si} - \frac{kT}{e_0}A\sqrt{C}.\tag{3.20}$$

And taking into account the contribution of C_Q^+ one will obtain

$$n = \frac{2kT}{Si} - \frac{kT}{e_0}A\sqrt{C}.\tag{3.21}$$

This approach was discussed in Adamson's monograph [2]. It was ascertained that the logic of equation (3.21) is justified only in the case when the ions of subphase are able to penetrate directly to the interface, that is, into the space between ionized monolayer molecules.

A lack of (3.21) deriving is the neglect of possible changes of activity coefficients of monolayer molecules and ions in subphase at different packing densities of monolayers and different ionic strength of subphase.

Davies approach

Another approach was proposed by Davies [8–9,19–20]. It concerns the case when the charges of monolayer molecules can be represented as regularly distributing on the polar group plane, and the distribution of subphase counter-ions takes place in the space of the diffusion part of double electrical layer. In so doing, electrostatic work on conversion from the system of uncharged surface + normal solution of uncharged particles to the charged surface + diffusion layer of opposite-charged electrolyte ions of subphase is calculated. And the surface pressure is determined as a derivative of free energy with respect to the monolayer area.

In deriving this approach the task is broken into two parts:

1. the description of energy of actually electrostatic field (ΔF_1)
2. the description of energy consumption, related to change of local concentration of subphase ions in a double electrical layer (ΔF_2), that is, osmotic component

The dependence of electrostatic potential φ_0 upon the surface charge density in this model is as follows:

$$\varphi = \frac{2kT}{e_0} Arsh \frac{e_0 Z}{S_i \sqrt{8CNkT\varepsilon_0\varepsilon_1}} = \frac{2kT}{e_0} Arsh \frac{e_0 Z}{S_i 2A\sqrt{C}}. \tag{3.22}$$

And the potential gradient for any section of a double electrical layer is described by the expression

$$\frac{d\varphi}{dz} = \frac{2A\sqrt{C}}{\varepsilon_0\varepsilon_1} sh \frac{e_0\varphi_z}{2kT}. \tag{3.23}$$

At so doing the component ΔF_1 can be calculated as convertible work of electrostatic field, which was done during the monolayer formation from molecules with constant charge (i.e., by compression of monolayer from the state with indefinitely small charge density up to a given one [19]):

$$\Delta F_1 = -\frac{\varepsilon_0\varepsilon_1}{2} \int\limits_{\infty}^{S_i} \int\limits_{\infty}^{0} (E)_z^2 dz dS, \tag{3.24}$$

where z – the direction along the normal from interface to subphase bulk, $E_z = -d\varphi/dz$ – electrostatic field strength along «z» direction, S_i – the area per ionized molecule in monolayer. Writing down (3.24) as

$$\Delta F_1 = -\frac{\varepsilon_0 \varepsilon_1}{2} \int\limits_{\infty}^{S_i} \int\limits_0^{\varphi_0} E_\varphi^2 d\varphi dS \qquad (3.25)$$

after the first integration one will receive:

$$\Delta F_1 = \int\limits_{\infty}^{S_i} \frac{2kT}{e_0} A\sqrt{C}\left[ch\frac{e_0\varphi_0}{2kT} - 1 \right] dS, \qquad (3.26)$$

where φ_0 is determined by equation (3.22).

After integration of (3.26) over a domain from ∞ to S_i one will get

$$\Delta F_1 = zkTArsh\frac{e_0 z}{S_i 2A\sqrt{C}} - \frac{2kT}{e_0} A\sqrt{C}\left[chArsh\frac{e_0 z}{S_i 2A\sqrt{C}} - 1 \right] \cdot S_i. \quad (3.27)$$

This expression describes actually electrostatic part of the Helmholtz free surface energy of system with the charge regularly distributed on its surface.

At calculating the osmotic component, the Boltzmann distribution of surface-inactive ions' concentration in diffusion part of double electrical layer is taken into account, that is:

$$C_s^+ = C^+ \exp\left(-\frac{e\varphi}{kT} \right)$$

$$C_s^- = C^- \exp\left(\frac{e\varphi}{kT} \right) \qquad (3.28)$$

where φ – a current meaning of Guoy–Chapman potential in subphase; C^+ and C^- – concentrations of surface-inactive ions. They equal to volume concentration of electrolyte C, if it indicates as type 1–1.

The surplus concentration of these ions in diffuse part of double electrical layer is equal to the difference between local concentrations and ion concentrations in subphase, that is:

$$C_{surpl} = C\left[\exp\left(-\frac{e_0\varphi}{kT} \right) + \exp\left(\frac{e_0\varphi}{kT} \right) - C \right] = 2C\left[ch\frac{e_0\varphi}{kT} - 1 \right]. \qquad (3.29)$$

Taking into account (3.29), the elementary change of free energy due to the appearance of ions' surplus in the field of action of monolayer charges is described by the ratio

$$\Delta F_2 = \int\limits_{\infty}^{S_i} \int\limits_{\infty}^{0} 2CNkT \left(ch\frac{e_0\varphi}{kT} \right) dz dS. \tag{3.30}$$

The internal integral in (3.30) displays ion distribution effect through the depth of subphase. The external integral shows that ΔF_2 concerns to the process of monolayer formation from molecules with constant charge (method of compression of monolayer from indefinitely small charge density up to given).

The first integration of (3.30) is made by substitution of z for $\varphi(dz = -d\varphi/E)$. Thus, taking into account (3.23), the equation (3.30) in parameters φ and S can be written in the following form:

$$+\Delta F_2 = -\int\limits_{\infty}^{S_i} \int\limits_{0}^{\varphi} \frac{1}{2} A\sqrt{C} \cdot sh\frac{e_0\varphi}{2kT} d\varphi dS. \tag{3.31}$$

Accordingly, intermediate decision takes the form

$$+\Delta F_2 = \int\limits_{\infty}^{S_i} 2A\sqrt{C} \left(\frac{kT}{e_0} \right) \cdot \left[ch\frac{e_0\varphi}{2kT} - 1 \right] \cdot dS. \tag{3.32}$$

The final decision is obtained by integration of (3.32) over S, taking into account (3.22):

$$\Delta F_2 = zkTArsh\frac{e_0 z}{S_i 2A\sqrt{C}} - \left(\frac{kT}{e_0} \right) 2A\sqrt{C}\left[chArsh\frac{e_0 z}{S_i 2A\sqrt{C}} - 1 \right] \cdot S_i. \tag{3.33}$$

The expressions (3.33) and (3.27) are identical analytically; therefore, the general value of Helmholtz free energy for monolayer with uniformly smeared charge is equal to

$$\Delta F = 2zkTArsh\frac{e_0 z}{S_i 2A\sqrt{C}} - \left(\frac{2kT}{e_0} \right) 2A\sqrt{C}\left[chArsh\frac{e_0 z}{S_i 2A\sqrt{C}} - 1 \right] \cdot S_i. \tag{3.34}$$

The same result will be obtained if we shall take the integral for hypothetical monolayer charging process, assuming that charge density in it grows from zero up to σ at constant area

$$\Delta F = \Omega \int_{0}^{\sigma} \varphi d\sigma.$$ (3.35)

As a result the electrostatic component of surface pressure in ionized monolayer can be obtained by differentiation of ΔF with respect to S_i:

$$\Pi_{el} = -\frac{d}{dS_i}\left[2zkTArsh\frac{e_0 z}{S_i 2A\sqrt{C}} - \left(\frac{2kT}{e_0}\right)2A\sqrt{C}\left(chArsh\frac{e_0 z}{S_i 2A\sqrt{C}} - 1\right)\cdot S_i\right]$$

$$= \left(\frac{2kT}{e_0}\right)2A\sqrt{C}\left(chArsh\frac{e_0 z}{S_i 2A\sqrt{C}} - 1\right).$$ (3.36)

The result coinciding with (3.36) can be also achieved by the other way, namely, by realizing the hypothetical process of charging of monolayer with constant area, where the surface potential is integration parameter:

$$\Pi_{el} = -\frac{d}{ds}\Delta F = \int_{0}^{\varphi} \sigma d\varphi.$$ (3.37)

Hence, the surface pressure by Davies is equivalent to Gibbs free energy loss associated with double electrical layer formation in subphase under uniformly charged interface [2].

Limit relations for surface pressure in Donnan and Davies models

In the case of low electrolyte concentration in subphase, that is, when the value $e_0 z / S_i \sqrt{8CNkT\varepsilon_0 \varepsilon_1}$ is large, the expression for Π_{el} according to Davies has the following form:

$$\Pi_{el} = \frac{2zkT}{S_i} - \frac{2kT}{e_0}2A\sqrt{C}.$$ (3.38)

Respectively, the total surface pressure of ionized monolayer (taking into account kinetic component) should be equal:

$$\Pi = \frac{kT}{S_i} + \frac{2zkT}{S_i} - \frac{2kT}{e_0}2A\sqrt{C}.$$ (3.39)

Or, for monolayer consisting of singly charged surface-active ions:

$$\Pi = \frac{3kT}{S_i} - \frac{2kT}{e_0}2A\sqrt{C}.$$ (3.40)

At low ionic strength of subphase ($I \sim 10^{-6} \div 10^{-4}$) the absolute value of $2A\sqrt{C}$ is essentially lower than $2kT/e_0$. And thus, the surface pressure in monolayers formed from 1–1 charged surface-active substances (SAS) on subphases with low ionic strength should be described by the following equation:

$$\Pi \approx \frac{3kT}{S_i}. \tag{3.41}$$

In Donnan model (equation 3.16) the expression for surface pressure under similar conditions is the following:

$$\Pi = \frac{2kT}{S_i}. \tag{3.42}$$

In the other extreme case, namely at big ionic strength of subphase, when

$$Arsh\left(\frac{e_0 z}{S_i 2A\sqrt{C}}\right) \approx \frac{e_0 z}{S_i 2A\sqrt{C}}, \tag{3.43}$$

the expression for Π_{el} in Davies model is transformed to the following form:

$$\Pi_{el} \cong \frac{2kT}{e_0} 2A\sqrt{C}\left[ch\frac{e_0 z}{S_i 2A\sqrt{C}} - 1\right] \approx \frac{1}{2} \cdot \frac{(e_0 z)^2}{\chi S_i^2 \epsilon_0 \epsilon_1}. \tag{3.44}$$

It is easy to see this at chx expanding into series:

$$chx = 1 + \frac{x^2}{2!} + \frac{x^4}{4!} + \frac{x^6}{6!} + ..., \tag{3.45}$$

and at restricting to the sum of two first members:

$$\Pi_{el} \cong \frac{2kT}{e_0}\sqrt{8CNkT\epsilon_0\epsilon_1} \cdot \frac{1}{2}\left(\frac{e_0 z}{S_i\sqrt{8CNkT\epsilon_0\epsilon_1}}\right)^2$$

$$\approx \frac{(e_0 z)^2 kT}{S_i^2 \sqrt{8CNkT\epsilon_0\epsilon_1}} \approx \frac{(e_0 z)^2}{2\chi\epsilon_0\epsilon_1 S_i^2} \approx \frac{1}{2} \cdot \frac{e_0 z \varphi_0}{S_i}. \tag{3.46}$$

The ratio (3.46) represents the known Helmholtz formula for a plane condenser model with fixed distance between condenser plates (which is equal to Debay shielding length).

A similar expression for Π_{el} in the Donnan model with allowance for big ionic strength of subphase electrolyte is obtained by transformation of the right part of the equation (3.15):

$$\Pi_{el} \cong \frac{kT}{e_0}\sqrt{2CNkT\varepsilon_0\varepsilon_1} \cdot \frac{1}{2}\left(\frac{2e_0 z}{S_i\sqrt{8CNkT\varepsilon_0\varepsilon_1}}\right)^2$$

$$\approx \frac{e_0 z^2 kT}{S_i^2\sqrt{8CNkT\varepsilon_0\varepsilon_1}} \approx \frac{1}{2}\cdot\frac{e_0 z\varphi_0}{S_i} \approx \frac{1}{2}\cdot\frac{q\varphi_0}{S_i}. \tag{3.47}$$

From (3.46) and (3.47) it is possible to conclude that the electrostatic component of surface pressure in monolayers formed from singly charged molecules on subphases with high ionic strength will be equal to the kinetic component if:

$$\frac{e}{\sqrt{8CNkT\varepsilon_0\varepsilon_1}} = S_i\sqrt{C}. \tag{3.48}$$

But when $S_i\sqrt{C}$ is essentially greater than $e/\sqrt{8CNkT\varepsilon_0\varepsilon_1}$, the electrostatic component of surface pressure should tend to zero and surface pressure will arise from thermal motion of monolayer molecules. Thus, the Donnan and Davies models predict the values of surface pressure within the limits from

$$\Pi = \frac{kT}{S_i}\text{(high ionic strength of subphase)} \tag{3.49}$$

up to

$$\Pi = \frac{2kT}{S_i} \quad \text{or} \quad \Pi = \frac{3kT}{S_i}\text{(low ionic strength of subphase)} \tag{3.50}$$

Surface pressure by Phillips and Rideal model

Phillips and Rideal [10,11] have proposed to analyze the surface pressure in ionized monolayers on the basis of equation:

$$\Pi = \frac{2kT}{S_i} + 2A\sqrt{C}\frac{2kT}{e_0}\cdot chArsh\left[\frac{e_0 z}{S_i 2A\sqrt{C}} - 1\right]. \tag{3.51}$$

This equation is different from the Davies equation by a factor of «2» in kinetic term of the right part. The logic was based on the analysis of fundamental ratio:

$$\Pi = E_R + \frac{\Omega d E_R}{d\Omega} - E_{disp} - \frac{\Omega d E_{disp}}{d\Omega} + \frac{kTsC_s^{R^+}}{C_s^{R^+}d\Omega} + \frac{kTsC_s^{x^-}}{C_s^{x^-}d\Omega}, \qquad (3.52)$$

where Π – surface pressure in monolayer, E_R and E_{disp} – energy of electrostatic repulsion and molecular attraction of monolayer molecules per interface unit, $C_s^{R^+}$ and $C_s^{x^-}$ – molar fraction of surface-active ions and opposite ions, Ω – the surface area, and S – specific area.

Assuming that

$$E_R = \frac{e_0 \varphi_0}{\Omega} - F, \qquad (3.53)$$

where

$$F = \frac{2kT}{e_0} 2A\sqrt{C} \cdot \left[chArsh \frac{e_0 z}{S_i 2A\sqrt{C}} - 1 \right], \qquad (3.54)$$

Phillips and Rideal have received the Davies expression for the sum of the second and third components in (3.52):

$$E_R + \frac{\Omega d E_R}{d\Omega} = \frac{2kT}{e_0} 2A\sqrt{C} \cdot \left[chArsh \frac{e_0 z}{S_i 2A\sqrt{C}} - 1 \right]. \qquad (3.55)$$

As long as the dispersion component in monolayers with very low and moderate packing density of molecules is small, that is, $F_{disp} + \frac{\Omega d E_{disp}}{d\Omega} \to 0$ at $S_i > 1$ nm^2, Phillips and Rideal [11] have estimated the contribution of the last two members in equation (3.52) and have obtained kT doubling in kinetic component (3.51). As argument confirming the correctness of (3.51), the authors have attracted the analogy with formation of osmotic pressure in solutions of 1–1 -valence electrolytes, where the following ratio is true:

$$P_{osm} \cdot V = 2RT, \qquad (3.56)$$

instead of $P_{osm} \cdot V = RT$ for nonelectrolyte solution.

Thus, according to Phillips and Rideal [10,11] the surface pressure in ionized monolayers should be followed to the equation:

$$\Pi \approx \frac{4kT}{S_i}. \qquad (3.57)$$

Phillips and Rideal have compared the values of surface pressure in monolayers formed from sodium dodecylsulphate and sodium octadecylsulphate, calculated on (3.51). The course of experimentally measured surface pressure was consistent with calculated results for monolayers from dodecylsulphate in the interval of specific areas 50÷4 nm²/molecule.

In octadecylsulphate monolayers the measured value of surface pressure was smaller than it follows from (3.51). In this connection, it was suggested to make a correction in (3.51) for possible effects of monolayer molecules association, that is, to write down the equation (3.51) as:

$$\Pi = (2 - \alpha)\frac{kT}{S_i} + \frac{2kT}{e_0} 2A\sqrt{C} \cdot \left[chArsh\frac{e_0 z}{S_i 2A\sqrt{C}} - 1 \right] \qquad (3.58)$$

where α – factor considering association.

Selecting magnitudes of α empirically, it was possible to receive the good agreement of calculated values of Π with the shape of experimental compression isotherms. It is in order to note that in spite of the satisfactory agreement between calculated data upon (3.58) and (3.51) and experimental ones, the correctness of physical sense of these equations raises doubts, because in deriving these expressions the osmotic component contribution was considered two times. At the same time the coincidence of really generated surface pressure in the ionized monolayers of sodium dodecylsulphate calculated upon (3.51), which have a limit

$$\Pi \to \frac{4kT}{S_i}$$

shows that the level of $3kT/S_i$ can be exceeded depending upon the nature of monolayer-forming molecules and ion concentration of a subphase.

It is reasonable to connect this with the fact that the surface pressure in real monolayers is stipulated by interaction of a lot of discrete molecules and surface-active ions. They have the final sizes, and fall close to the phase with low dielectric permeability. In a surface layer the strength of interaction between ionized molecules essentially depends upon the manner of these molecules' packing, upon their charges, and upon concentration of subphase electrolyte and dielectric permeability in the boundary layer.

The model of Bell, Levine, and Pethica

The effects concerned with discrete structure of monolayers were analyzed in works of Bell, Levine, and Pethica (BLP) [3,13,15–19]. These authors estimated the contribution of the so-called «fluctual» component, which is caused by distinction between the real potentials in location sites of separate monolayer molecules and the average potentials, received following the Guoy–Chapman model. Simultaneously considered was that the dielectric permeability in localization plane of ionized monolayer molecules is much smaller than the dielectric permeability of subphase bulk.

The general course of discussion while calculating the fluctual component consists in the following. Bell, Levine, and Pethica analyzed a monolayer with area Ω, containing N surface-active molecules spreading at water. The monolayer volume was assumed as negligibly small in comparison with the subphase volume, and all processes were considered as isothermal. The authors also have assumed that the monolayer surface gets a charge as a result of dissociation of N_i molecules. The number of surface-active ions (n_i) and general number of surface-active molecules per square centimeter of monolayer surface (n) was believed equal to

$$n_i = \frac{N_i}{\Omega}; \quad n = \frac{N}{\Omega}. \tag{3.59}$$

Effective density of charges in monolayer was calculated as a product of the number of dissociated molecules for value of their charge:

$$\sigma = \pm e n_i. \tag{3.60}$$

Chemical potentials of undissociated monolayer molecules (μ_u), surface-active ions (μ_i), and surface-inactive ions (μ_e) were connected with free monolayer energy by the ratio

$$\left(\frac{d\Delta F}{dn_i}\right)_{T,V,N} = \mu_i + \mu_e + \mu_u = \Delta\mu, \tag{3.61}$$

where μ_i – the real difference between «fluctual» (μ_{ic}) and continual $(e\varphi)$ components of chemical potential of surface-active ions:

$$\mu_i = \mu_{ic} - e\varphi. \tag{3.62}$$

By definition of BLP, fluctuation part reflects a difference between really existing potential on surface-active ion and average potential, calculated according to the Guoy–Chapman theory. Thus the control of discrete monolayer structure was made by means of μ_{ic}.

When the concentration of subphase electrolyte is much greater than the concentration of surface-active ions, under equilibrium conditions the value of μ_e is constant and does not depend upon n_i. The value of chemical potential of nondissociated molecules (μ_u) was also accepted as independent upon n_i. As free energy of ionized monolayer represents a difference between free energy of interface filled by ionized molecules (the total number of molecules on the surface – n, the number of ionized molecules $- n_i$) and free energy of interface containing uncharged mono-layer with the same amount of molecules $- n$ (but in this case $n_i = 0$, and dissociation degree of monolayer molecules $\alpha = n_i/n = 0$), the general expression for free energy of ionized monolayer was submitted as [18]:

$$\Delta F_i = \int_0^{n_i} \Delta\mu_i dn_i. \tag{3.63}$$

Or, taking into account (3.61)–(3.62), and after substitution of dn_i for $nd\alpha$, BLP have received

$$\Delta F_i = n\int_0^\alpha \Delta\mu_{ic}d\alpha + \int_0^\alpha \varphi d\sigma = \Delta F_{ic} + \Delta F_{el}, \tag{3.64}$$

where $d\alpha = \frac{1}{n}dn$, ΔF_{ic} – the fluctuation component of free energy, ΔF_{el} – the continual component of free energy.

The expression (3.63) is related to free energy of monolayer per unit area. Multiplication of (3.63) by Ω and differentiation with respect to area gives:

$$\Delta n_{el} = \frac{-d}{d\Omega}\left(\Omega\Delta F_{ic}\right) - \frac{d}{d\Omega}\left(\Omega\Delta F_{el}\right) = \Delta n_{ic} + \Delta nel. \tag{3.65}$$

The first term in the right part of (3.65) was transformed to the formula suitable for analysis, as follows [18]:

$$\Delta n_{ic} = -\frac{d}{d\Omega}\left(\Omega\Delta F_{ic}\right) = -\frac{d}{d\Omega}N\int_0^\alpha \Delta\mu_{ic}d\alpha$$

$$\tag{3.66}$$

$$= -N\int_0^\alpha \left(\frac{d\Delta\mu_{ic}}{dn}\right)_\alpha d\alpha \frac{dn}{d\Omega} = n^2\int_0^\alpha \left(\frac{d\Delta\mu_{ic}}{dn}\right)_\alpha d\alpha$$

where $\dfrac{N}{n} = \Omega$, $\dfrac{dn}{d\Omega} = -\dfrac{n}{\Omega}$.

The analytical expression for the $\Delta\Pi_{el}$ component was obtained identically to the Davies one. Thus, the calculation of electrostatic component of surface pressure by the BLP model assumes calculation of two components:

$$\Pi_{el} = \int_0^\sigma \sigma d\varphi + n^2\int_0^\alpha \left(\frac{d\Delta\mu_{ic}}{dn}\right)_\alpha d\alpha. \tag{3.67}$$

From (3.67) it is possible to conclude that the Davies model for calculation of the electrostatic component of surface pressure is acceptable, if the value $(d\Delta\mu_{ic}/dn)$ is equal to zero. If the value $(d\Delta\mu_{ic}/dn)$ deflects from zero to negative or positive sides, the results will be as calculated upon Donnan, or Phillips–Rideal models.

Bell, Levine, and Pethica have applied the equation (3.67) to estimate the effects of charge discreteness in monolayers formed from 1–1 charged SAS. Toward this end they have calculated "fluctual" potential Φ, that is, the potential of "own atmosphere" of surface-active ions of monolayer [16–18]. It gives the correction for the difference between Guoy–Chapman potential and "real" potential (micro potential) in location sites of each monolayer molecule. At calculating fluctuation potential, it was considered that the charges of real monolayer molecules are concentrated in the very limited space. In so doing, Grem's approach was taken, and calculation was made by means of the cut-out disk method. According to this method the surface-active ions can be determined to be located on adsorption plane in the centers of the circles with radius r_0. The regularly smeared charges were believed withdrawn from the abovementioned circles. The value r_0 was associated with average charge density in monolayer by ratio:

$$\pi r_0^2 \sigma = -e, \tag{3.68}$$

where $-e$ – a charge of arbitrary chosen «central» surface-active ion.

As the fluctual potential essentially depends upon position of the adsorption plane with respect to the interface, in the case of aqueous electrolyte solution–monolayer–air system, the authors have made three assumptions:

1. The ionized parts of monolayer molecules are located in a plane parallel to subphase surface and displaced from it at fixed distance β.
2. Above the interface the medium dielectric permeability has a constant value ε, and under the interface, $-\varepsilon_p$.
3. Subphase electrolyte is so diluted that its ions practically do not shield electrostatic interactions between the ionized monolayer molecules.

Taking into account these assumptions, Bell, Levine, and Pethica have received the following expression for fluctual potential:

$$\Phi = \pm \frac{\sigma}{\varepsilon} \int_0^{r_0} \left[\frac{1}{r} + \frac{\varepsilon - \varepsilon_p}{\varepsilon + \varepsilon_p} \frac{1}{\sqrt{r^2 + 4\beta^2}} \right] r\, dr = \pm \frac{\sigma}{2\varepsilon} \left(r_0 + \frac{\varepsilon - \varepsilon_p}{\varepsilon + \varepsilon_p} \sqrt{r_0^2 + 4\beta^2} - 2\beta \right), \tag{3.69}$$

where σ – average density of surfaces charge in monolayer, and (+) – corresponds to a case of surface-active cations. The second component of underintegral expression of equation (3.69) evaluates the contribution of reflection forces at the interface.

The contribution of Φ to $\Delta\mu_{ie}$ value was found on the basis of ideas about Guntelberg's charging process. The charge e on the separate surface-active ion was believed to vary from zero up to finite quantity e_0, while the charge density was constant outside of cut-out disk. Thus, according to (3.68) and (3.69) Bell, Levine, and Pethica have received:

$$\Delta\mu_{ic} = -\int_0^{e_0} \Phi de = 2\pi\sigma \int_0^{r_0} \Phi_{(r_0)} r_0 dr_0$$

$$= \frac{\pi\sigma^2}{\varepsilon}\left\{\frac{r_0^3}{3} + \frac{\varepsilon - \varepsilon_p}{\varepsilon + \varepsilon_p}\left[\frac{1}{3}\left(r_0^2 + 4\beta^2\right)^{3/2} - \frac{8\beta^3}{3} - \beta r_0^2\right]\right\}.$$

(3.70)

As follows from (3.70) the value $\Delta\mu_{ic}$ depends upon the α only over the surface charge density $\sigma = -\alpha n e$. Therefore according to (3.66) an expression for $\Delta\Pi_{ic}$ was transformed as follows:

$$\Delta n_{ic} = n^2 \int_0^\alpha \left(\frac{d\Delta\mu_{ic}}{dn}\right)d\alpha = n^2 \int_0^\alpha \frac{d\Delta\mu_{ic}}{d\sigma}\left(\frac{d\sigma}{dn}\right)d\alpha$$

$$= \frac{d\Delta\mu_{ic}}{e} + \frac{1}{e}\int_0^\alpha \Delta\mu_{ic}d\sigma = \frac{\Delta\mu_{ic}}{\pi r_0^2} - \frac{2}{\pi}\int_{r_0}^\infty \frac{\Delta\mu_{ic}}{r_0^3}dr.$$

(3.71)

This transfer is based on application of the second charging process, when the charge density in monolayer grows from zero up to finite quantity of σ. Substituting the analytical expression for $\Delta\mu_{ic}$ from (3.71) to the formula:

$$\frac{2}{\pi}\int_{r_0}^\infty \frac{\Delta\mu_{ic}}{r_0^3}dr_0$$

(3.72)

authors of the BLP model have received:

$$-\frac{2}{\pi}\int_{r_0}^\infty \Delta\mu_{ic}\frac{dr_0}{r_0} = \frac{2e^2}{\varepsilon\pi^2}\left\{\frac{1}{9r_0^3} + \frac{\varepsilon - \varepsilon_p}{\varepsilon + \varepsilon_p}\left[\begin{array}{c}\left(\dfrac{2\beta^2}{9r_0^6} + \dfrac{7}{72r_0^4} + \dfrac{1}{192r_0^2\beta^2}\right)\sqrt{r_0^2 + 4\beta^2} \\[3mm] -\dfrac{1}{384\beta^3}\ln\dfrac{2\beta + \sqrt{4\beta^2 + r_0^2}}{r_0} - \dfrac{4\beta^3}{9r_0^6} - \dfrac{1\beta}{4r_0^4}\end{array}\right]\right\}$$

(3.73)

Accordingly, the final expression for fluctual member (with allowance for assumptions 1–3) has got a kind:

$$\Delta n_{ic} = -\frac{e^2}{\varepsilon\pi^2}\left[\frac{1}{9r_0^3}+\frac{\varepsilon-\varepsilon_p}{\varepsilon+\varepsilon_p}f(r_0,2\beta)\right],$$

where

$$f(r_0,2\beta)=\left(\frac{8\beta^2}{9r_0^6}+\frac{5}{36r_0^4}-\frac{1}{96r_0^2\beta^2}\right)\sqrt{r_0^2+4\beta^2}$$

$$+\frac{1}{192\beta^3}\ln\frac{2\beta+\sqrt{4\beta^2+r_0^2}}{r_0}-\frac{\beta}{2r_0^4}-\frac{16\beta^3}{9r_0^6}.$$

(3.74)

Then the authors of equation (3.74) have ascertained that if assumption 3 is not fulfilled, the correction of effect of charges screening of monolayer forming molecules by subphase ions should be made by location of two additional disks with radius r_0 and charge density σ at a distance $(\chi^{-1}+2\beta)$ from geometrical interface. Thus (3.69) was transformed to a kind:

$$\Phi=\frac{\sigma}{2\varepsilon}\int_0^{r_0}\left[\frac{1}{\sqrt{r_0^2+(\chi^{-1})^2}}+\frac{\varepsilon-\varepsilon_p}{\varepsilon+\varepsilon_p}\frac{1}{\sqrt{r_0^2+\chi^{-1}+2\beta^2}}\right]r_0 dr_0$$

$$=\frac{\sigma}{2\varepsilon}\left\{\sqrt{r_0^2+(\chi^{-1})^2}-\chi^{-1}+\frac{\varepsilon-\varepsilon_p}{\varepsilon+\varepsilon_p}\left[\sqrt{r_0^2+(\chi^{-1}+2\beta)^2}-(\chi^{-1}+2\beta)\right]\right\}.$$

(3.75)

Repeating the calculating procedure on the basis of (3.70) to (3.74) Bell, Levine, and Pethica have received:

$$\Delta n_{ic}=\frac{e^2}{\varepsilon\pi^2}\left\{f(r_0,\chi^{-1})-\frac{1}{9r_0^3}+\frac{\varepsilon-\varepsilon_p}{\varepsilon+\varepsilon_p}\left[f(r_0,\chi^{-1}+2\beta)-f(r_0,2\beta)\right]\right\}. \quad (3.76)$$

Equation (3.76) describes $\Delta\Pi_{ic}$ in the system with discrete package of monolayer-forming molecules, which are placed in adsorption plane at a distance β from interface. In the plane of adsorption and above, its dielectric permeability is equal to ε. In subphase bulk it takes the value ε_p. The opposite charges of subphase electrolyte and their equivalent images in air are considered to locate at fixed distances from adsorption plane, equal to $(\chi^{-1}+2\beta)$ and (χ^{-1}), accordingly.

Taking into account (3.76), the final expression for surface pressure in ionized monolayer has appeared as follows:

$$n = \frac{kT}{S_i} + \int\limits_0^\sigma \sigma d\varphi + \frac{e^2}{\varepsilon\pi^2}\left(f\left(r_0, \chi^{-1}\right) - \frac{1}{9r_0^3} + \frac{\varepsilon - \varepsilon_p}{\varepsilon + \varepsilon_p}\{f\left[r_0, \left(\chi^{-1} + 2\beta\right)\right] - f\left(r_0, 2\beta\right)\}\right)$$

$$(3.77)$$

However, the equation (3.77) has two essential lacks. The first is that there was an accepted condition of fixed distance between condenser plates in the diffusion part of double electrical layer. Equivalent opposite charge in subphase and its image in air medium is believed to locate at a distance $(\chi^{-1} + 2\beta)$ from geometrical interface, irrespective of the charge value of monolayer-forming molecules. This condition can be observed only for monolayers with low surface-charge density, when the structure of double electrical layer in subphase is similar to Helmholtz's. The second lack is the ignoring of the dimensions of surface-active ions in monolayer. This is allowable only under limited conditions.

The certain contribution to the generation of electrostatic component of surface pressure also can be made by effects caused by local curvatures of subphase surface in the location sites of separate monolayer molecules. The authors of the BLP model didn't take into account these circumstances.

Calculation of surface pressure on the basis of statistical mechanics approach

Surface pressure in any monolayer can be calculated on the base of statistical mechanics if one considers monolayer as a two-dimensional ensemble of chemically identical molecules. In this case the following equation is valid [7]:

$$\sum \frac{1}{2}mc^2 = -\frac{1}{2}\sum (Xx + Yy).$$

$$(3.78)$$

The left part of (3.78) represents the sum of molecules of kinetic energy in two-dimensional ensemble, and the right, the sum of interaction energy of all molecules pairs. For each molecule, $\frac{1}{2}mc^2 = \frac{1}{2}mu^2 + \frac{1}{2}mv^2$.

When the system is subject to the laws of classical mechanics, the requirement for uniform distribution of molecules' energy upon freedom degrees results in expression:

$$NkT = -\frac{1}{2}\sum(Xx+Yy), \tag{3.79}$$

where N – total number of molecules in examined system.

The right part of the equation (3.79) represents complex value. It includes the components conforming with external and internal forces, that is:

$$NkT = -\frac{1}{2}\sum(Xx+Yy)_{ex} - \frac{1}{2}\sum(Xx-Yy)_{in} \tag{3.80}$$

To estimate the first member of this equation, it is necessary to analyze a rectangle on the surface with length x and width y. If the surface pressure inside this rectangle is equal to Π, then the molecules of surface ensemble will exert pressure upon the right wall with a force Π_y, and the wall will counteract with a force $-\Pi_y$. An average distance between molecules and the right wall is equal to $\frac{1}{2}x$, so the average value $\sum Xx$ is equal to $-\frac{1}{2}\Pi yx$. The average value $\sum Yy$ for the lower and top walls is found similarly. It is equal to $-\frac{1}{2}\Pi yx$. As $xy = \Omega$, hence:

$$-\frac{1}{2}\sum(Xx+Yy)_{ex} = \Pi_s\Omega, \tag{3.81}$$

and

$$NkT = \Pi_s\Omega - \frac{1}{2}\sum(Xx+Yy)_{in}, \tag{3.82}$$

or

$$\Pi_s\Omega = NkT + \frac{1}{2}\sum(Xx+Yy)_{in}, \tag{3.83}$$

where Ω – the total area of surface ensemble.

Surface pressure in non-ionized monolayers

Let's examine a monolayer consisting of non-ionized amphiphilic molecules. We shall believe that these molecules do not have dipoles, so they are indifferent relative to subphase electrolyte. Accordingly, the double electrical layer in subphase is absent. In such a way, only dispersion and kinetic forces can affect between monolayer molecules. For simplicity we will believe that monolayer is extended, and therefore the dispersion forces are negligibly small. Eventually the surface pressure in monolayer will form only at the cost of thermal motion. So, the components of total forces (X and Y), affecting on each molecule in monolayer, are subject to the second law of Newton [7]:

$$X = m\frac{d^2x}{dt^2} \tag{3.84}$$

$$Y = m\frac{d^2y}{dt^2} \tag{3.85}$$

where x and y – coordinates of gravity centers of molecules with mass m in the surface ensemble.

Accordingly, the sum of products of values X and Y upon coordinates of any molecule (x, y) is equal to:

$$Xx + Yy = m\frac{d}{dt}\left(x\frac{dx}{dt} + y\frac{dy}{dt}\right) - m\left[\left(\frac{dx}{dt}\right)^2 - \left(\frac{dy}{dt}\right)^2\right]. \tag{3.86}$$

As far as $x\dfrac{dx}{dt} = \dfrac{1}{2}\cdot\dfrac{d}{dt}x^2$, and $(dx/dt)^2$ is the square of linear speed of any molecule in a given direction, the following equation is equivalent to ratio (3.86):

$$Xx + Yy = \frac{1}{2}m\frac{d}{dt}\left[\frac{d}{dt}(x^2 + y^2)\right] - mc^2 \tag{3.87}$$

where c^2 – the sum of squares of linear molecules rates at x and y directions.

In monolayer consisting of many molecules (N^*), it is possible to consider that statistic–mean position of each molecule is constant: $x^2 + y^2 = const$ [7]. In this connection, the value of the first term in the right part of equation (3.87) will be negligible small and the following equation will be true:

$$\frac{1}{2}\sum_1^{N^*}(Xx + Yy) = -\frac{1}{2}\sum_1^{N^*}mc^2 \tag{3.88}$$

The left part of this equation is a virial.

For the classical motion laws the theorem of virial proves that:

$$-\frac{1}{2}\sum_1^{N^*}(Xx + Yy) = \Pi\Omega \tag{3.89}$$

where Π – surface pressure and Ω – area of monolayer.

Or:

$$-\frac{1}{2}\sum_{1}^{N^*}(Xx+Yy)=N^*kT. \tag{3.90}$$

Thus:

$$\Pi\Omega = N^*kT \tag{3.91}$$

$$\Pi = \frac{N^*}{\Omega}kT = \frac{kT}{S_i} \tag{3.92}$$

where S_i – specific area in monolayer.

Surface pressure at the presence of charges in monolayer

If far-affecting forces are present in monolayer, for example, electrostatic forces, then their calculation according to (3.80) must be carried out by means of equation:

$$N^*kT = -\frac{1}{2}\sum_{1}^{N^*}(Xx+Yy)-\frac{1}{2}\sum_{1}^{N^*}(\tilde{X}x+\tilde{Y}y) \tag{3.93}$$

where $\tilde{X}$ and $\tilde{Y}$ – components of far-affecting forces of intermolecular interaction.

For any molecules pair in monolayer space the values $(\tilde{X}_i)$ are described as follows (see Figure 3.1):

$$\tilde{X}_1 = -k\cos\theta = -\frac{\tilde{k}(x_2-x_1)}{\ell} \tag{3.94}$$

$$\tilde{X}_2 = k\cos\theta = \frac{\tilde{k}(x_2-x_1)}{\ell} \tag{3.95}$$

where $\tilde{k}$ – force of interaction of any pair of monolayer molecules along line of their weight centers, ℓ – distance between the weight centers of these molecules, θ – angle between weight centers' line and direction, in which this force is calculated. The sum of products $\tilde{X}_i$ upon x for any molecules pair comes up to:

$$\sum\tilde{X}x = \frac{\tilde{k}}{\ell}(x_2-x_1)^2. \tag{3.96}$$

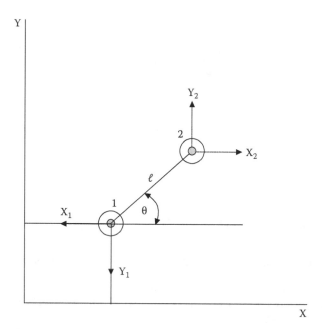

Figure 3.1 Schematic representation of partial forces between interacting species in two-dimensional ensemble.

At y direction the expression for the sum is similar:

$$\sum \tilde{Y}y = \frac{\tilde{k}}{\ell}(y_1 - y_2)^2.$$ (3.97)

Accordingly, for N^* molecules at the surface we shall receive:

$$\frac{N^*}{2} \sum_1^{N^*} (\tilde{X}x + \tilde{Y}y) = \frac{N^*}{2} \sum_1^{N^*} \left[\frac{\tilde{k}_i}{\ell_i}(x_1 - x_2)^2 + \frac{\tilde{k}_i}{\ell_i}(y_1 - y_2)^2 \right]$$

$$= \frac{N^*}{2} \sum_1^{N^*} \left(\frac{\tilde{k}_i}{\ell_i} \cdot \ell_i^2 \right) = \frac{N^*}{2} \sum_1^{N^*} \tilde{k}_i \ell_i$$ (3.98)

and as a result we shall get:

$$\Pi\Omega = N^*kT + \frac{N^*}{2} \sum^{N^*} \tilde{k}_i \ell_i.$$ (3.99)

The equation (3.99) describes the state of monolayer taking into account intermolecular interactions.

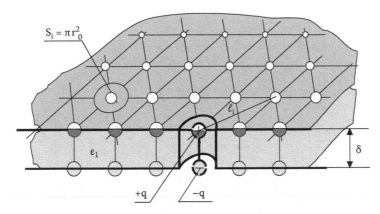

Figure 3.2 Schematic representation of ionized monolayer–subphase system in the case of discrete monolayer structure.

Since we are interested in the electrostatic component of surface pressure, let's represent the monolayer as discrete lattice with ionized molecules located in lattice points as shown in Figure 3.2. Let us assume that the polar heads of monolayer molecules have a spherical form and they are immersed partially to subphase with dielectric permeability ε_1. The charge of monolayer molecules is believed to be located on a surface of hemispheres, immersed into the space of subphase. This charge causes the appearance of a double electrical layer in subphase.

Without going into structural details of charge distribution in this system, it is possible to accept that the effective opposite charges of the diffusion part of double electrical layer are located at a distance δ from interface. In borders of this model the electrostatic force ($\tilde{k}_i$) of interaction between any molecules pair in monolayer space is determined by equation:

$$\tilde{k}_i = \frac{q^2}{4\pi\varepsilon_0\varepsilon_1} \cdot \left[\frac{1}{\ell_i^2} - \frac{1}{\ell_i^2 + \delta^2} \frac{\ell_i}{\sqrt{\ell_i^2 + \delta^2}} \right], \qquad (3.100)$$

where ℓ_i – the distance between the charge centers of interacting pairs of monolayer molecules, ε_1 – dielectric permeability of subphase, ε_0 – absolute dielectric permeability, and q – charge of monolayer molecules.

$q = ze$ – charge of ionized polar head of monolayer molecule,
  ℓ_i – distances between the weight centers of polar heads of monolayer molecules,
  δ – distance along the normal from the mass centers of ionized polar heads of monolayer molecules up to equivalent opposite charges in subphase,
$S_i = \pi r_0^2$ – specific area in a monolayer.

Or:

$$\tilde{k}_i = \frac{q^2}{4\pi\varepsilon_0\varepsilon_1}\left[\frac{1}{\ell_i^2} - \frac{1}{\ell_i^2+\delta^2}\frac{\ell_i}{\sqrt{\ell_i^2+\delta^2}}\right] = -\frac{dU}{d\ell}. \tag{3.101}$$

The substitution of (3.101) to (3.98) gives:

$$\frac{N^*}{2}\sum_1^{N^*}(\tilde{X}x+\tilde{Y}y) = \frac{N^*}{2}\sum_1^{N^*}\frac{q^2}{4\pi\varepsilon_0\varepsilon_1}\cdot\left[\frac{1}{\ell_i^2} - \frac{1}{\ell_i^2+\delta^2}\frac{\ell_i}{\sqrt{\ell_i^2+\delta^2}}\right]\cdot\ell_i. \tag{3.102}$$

For the monolayer formed from a large number of molecules, the summation in (3.102) can be replaced by integration. At that we accept that any of the chosen «central» ionized molecules and its opposite charge in subphase bulk is really discrete, and the charges of the other monolayer molecules and their opposite charges are regularly distributed on the appropriate planes. They are located at a distance r_0 from normal passing through the center of given molecule (Figure 3.3). So we record (3.102) as:

$$\frac{N^*}{2}\sum_1^{N^*}(\tilde{X}x+\tilde{Y}y) \cong \frac{N^*q^2}{2S_i\,4\pi\varepsilon_0\varepsilon_1}\cdot\int_{r_0}^{\infty}\left[\frac{1}{R^2} - \frac{1}{R^2+\delta^2}\frac{R}{\sqrt{R^2+\delta^2}}\right]_i 2\pi R^2 dR \tag{3.103}$$

where R – current distance.

After integration we shall receive:

$$\frac{N^*}{2\Omega}\sum_1^{N^*}(\tilde{X}x+\tilde{Y}y) \cong \frac{-q^2}{2\cdot2\varepsilon_0\varepsilon_1 S_i^2}\cdot\left[r_0 - \sqrt{r_0^2+\delta^2} - \frac{\delta^2}{\sqrt{r_0^2+\delta^2}}\right]. \tag{3.104}$$

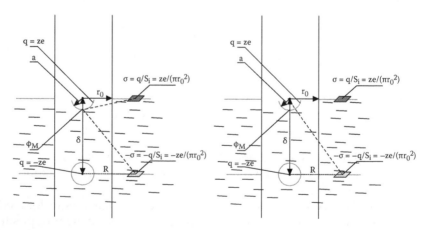

Figure 3.3 Schematic representation of elementary unit in the space of ionized monolayer with discrete packing of monolayer-forming molecules.

And as a result of substitution of (3.104) into (3.99) we shall get the expression for monolayer state taking into account the forces of electrostatic interaction between its molecules:

$$\Pi = \frac{KT}{S_i} + \frac{q^2}{4\varepsilon_0\varepsilon_1 S_i^2} \cdot \left[\sqrt{r_0^2 + \delta^2} + \frac{\delta^2}{\sqrt{r_0^2 + \delta^2}} - r_0 \right]. \tag{3.105}$$

The value δ, entering into (3.105), can be found if one knows the micro potential (φ_m) generated by all charges of system monolayer–subphase at the surface of a charged group of arbitrarily chosen (central) monolayer-forming molecules. As follows from Figure 3.3, the dependence of δ upon φ_m is

$$\varphi_m = \frac{q}{4\pi\varepsilon_0\varepsilon_1} \left\{ \frac{1}{a} - \frac{\delta}{\delta - a} + \frac{2\pi}{S_i} \left[\sqrt{r_0^2 + (\delta - a)^2} - \sqrt{r_0^2 + a^2} \right] \right\}, \tag{3.106}$$

where φ_m – micro potential, a – radius of ionized part of monolayer forming molecule, and S_i – specific area in monolayer.

So, we shall receive:

$$\varphi_m = \frac{ze}{4\pi\varepsilon_0\varepsilon_1} \left(\frac{1}{a} \right) - \frac{ze}{4\pi\varepsilon_0\varepsilon_1} \left(\frac{1}{\delta - a} \right) + \frac{ze}{S_i} \frac{2\pi}{4\pi\varepsilon_0\varepsilon_1} \int\limits_{r_0}^{\infty} \frac{R\,dR}{\sqrt{R^2 + a^2}}$$

$$- \frac{ze}{S_i} \frac{2\pi}{4\pi\varepsilon_0\varepsilon_1} \int\limits_{r_0}^{\infty} \frac{R\,dR}{\sqrt{R^2 + (\delta - a)^2}} \tag{3.107}$$

$$= \frac{ze}{4\pi\varepsilon_0\varepsilon_1} \left\{ \frac{1}{a} - \frac{1}{\delta - a} + \frac{2\pi}{S_i} \left[\sqrt{r_0^2 + (\delta - a)^2} - \sqrt{r_0^2 + a^2} \right] \right\}$$

and

$$\Pi_{s.c.dis.} = \frac{1}{2} \frac{N^*}{\Omega} \frac{ze}{4\pi\varepsilon_0\varepsilon_1} \int\limits_{r_0}^{\infty} \frac{ze2\pi R R\,dR}{S_i R^2} - \frac{1}{2} \frac{N^*}{\Omega} \frac{ze}{4\pi\varepsilon_0\varepsilon_1} \int\limits_{r_0}^{\infty} \frac{ze2\pi R R R\,dR}{S_i \sqrt{R^2 + \delta^2} \left(R^2 + \delta^2 \right)}$$

$$= \frac{1}{2} \frac{1}{S_i} \frac{(ze)^2}{2\varepsilon_0\varepsilon_1 S_i} \left[\int\limits_{r_0}^{\infty} dR - \int \frac{R^3 dR}{\left(R^2 + \delta^2 \right)^{3/2}} \right]$$

$$= \frac{1}{2} \frac{(ze)^2}{2\varepsilon_0\varepsilon_1 S_i^2} \left[\sqrt{r_0^2 + \delta^2} + \frac{\delta^2}{\sqrt{r_0^2 + \delta^2}} - r_0 \right]$$

$$= \frac{(ze)^2}{4\varepsilon_0\varepsilon_1 S_i^2} \left[\sqrt{r_0^2 + \delta^2} + \frac{\delta^2}{\sqrt{r_0^2 + \delta^2}} - r_0 \right]. \tag{3.108}$$

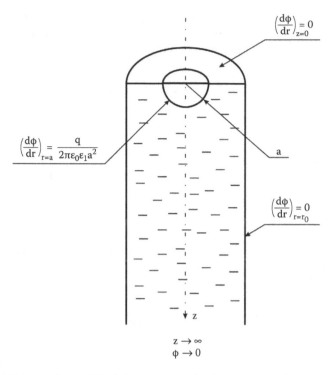

Figure 3.4 Schematic model of elementary cylinder as a separate unit of ionized monolayer with discrete packing and the boundary conditions in it.

The micro potential can be found by solving the Poisson–Boltzmann equation in tube domain (Figure 3.4), limited from above by area, equal to the specific area in monolayer (S_j) and outgoing to unlimited subphase depth. In this case the cylinder radius is equal to half of the average distance between monolayer-forming molecules.

The boundary conditions are as follows:

- the radial component of electrostatic field is equal to zero on the cylinder element due to symmetric packing of monolayer molecules
- the normal component of electrostatic field on the basis of cylinder (i.e., interface) is equal to zero everywhere except for the semi-spherical area occupied by a molecule

$$\left(\frac{d\varphi}{dz}\right)_{z=0} = 0. \tag{3.109}$$

This condition corresponds with total reflection of force lines of electrostatic field from polar–unpolar interface.

At the surface of head polar group of monolayer-forming molecule, the radial component of electrostatic field is determined by the following expression:

$$\left(\frac{d\varphi}{dr}\right)_{r=a} = \frac{q}{2\pi\varepsilon_0\varepsilon_1 a^2}. \tag{3.110}$$

Let's present the Poisson–Boltzmann equation in cylindrical coordinates:

$$\frac{d^2\varphi}{dr^2} + \frac{1 \cdot d\varphi}{r \cdot dr} + \frac{d^2\varphi}{dz^2} = -\frac{\rho}{\varepsilon_0\varepsilon_1}. \tag{3.111}$$

The equation (3.111) has the analytical solution at linear relation between potential φ and charge density (ρ). But it requires the numerical methods, when φ depends upon ρ nonlinearly.

The first case is equivalent to formation of monolayer from molecules with a small charge at subphase surface with high ionic strength. The second one is equivalent to formation of monolayer from molecules with a big charge on subphases with small ionic strength.

The influence of discrete charges upon the surface pressure in mono-layers is insufficiently studied. And the qualitative estimation of this influence is rather inconsistent. We have received numerical solutions of (3.111) in nonlinear variant and have compared the results of calculation of Π according to equation (3.105) with results calculated by known models and with experimental data. The description of calculating method is given in [21].

In connection with our task the calculated magnitudes of surface potentials of φ_m is given in Figure 3.5 in comparison with Guoy–Chapman potentials (φ_s). The charged hemispheres were used as the models of ionized groups of amphiphilic molecules.

As shown, the values of φ_m are always higher than the values of φ_s at identical charge densities in compared systems. This is due to specific electrostatic field distribution in the elementary cylinder and flat diffusion Guoy–Chapman layer. The plots of δ and Π_{scdis} versus packing density of at $C = 10^{-6}$ M and $z = 6$, calculated by the equations (3.104) and (3.105) with the use of found values of φ_m, are given in Figure 3.6.

Comparing this result with the data of surface pressure in melittin monolayer and with the results obtained in accord to the Davies and BLP models, it is easy to be convinced that under identical conditions (charge of monolayer molecules (z), their packing density at the interface (S_i), and electrolyte concentration of subphase (C)), the values of Π_{scdis} proved to be higher than the values of Π_{sc} predicted by the Davies and BLP models (Figure 3.7). The divergence becomes significantly higher as a charge of

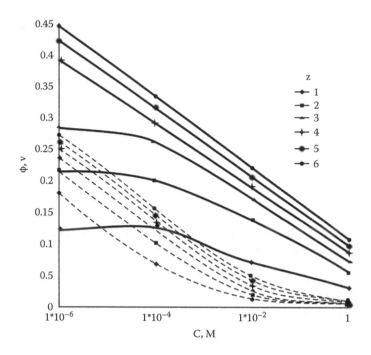

Figure 3.5 The calculated magnitudes of surface potentials (φ_m) versus subphase concentration at different values of molecules charges (z) obtained in accord with elementary cylinder model (solid lines) and the Davies model (dashed lines).

monolayer molecules increases and concentration of subphase electrolyte decreases. Evidently, the reason is that although the Davies model takes into account the effects of electrostatic energy in the diffusion part of double electrical layer, it does not represent the contribution of interaction forces of the ionized molecules inside the monolayer spaces as a two-dimensional ensemble.

This is confirmed by simple analysis of calculating procedure of ΔF_1 in accordance with Davies model. So, integration of equation (3.24) by means of substitution of dz for $-d\varphi/E$ gives a quite correct mathematical result, that is:

$$\Delta F_1 = -\frac{\varepsilon_0 \varepsilon_1}{2} \int\limits_{\infty}^{S_i} \int\limits_{\varphi_0}^{0} E_\varphi \, d\varphi \, ds = \int\limits_{\infty}^{S_i} \frac{kT}{e} 2A\sqrt{c} \left[ch\frac{e_0 \varphi_0}{2kT} - 1 \right] ds = \int\limits_{0}^{S_i} \int\limits_{0}^{\varphi} \sigma \, d\varphi \, dS. \quad (3.112)$$

But, according to the viewpoint of physics, this result is not exhaustive, since it does not take into account the action effect of the normal component of electrostatic field along the borders of the monolayer–subphase

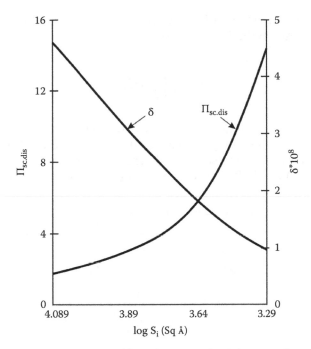

Figure 3.6 The calculated plots of δ and Π_{scdis} vs. $\log S_i$ for monolayer model with $z = 6$ and $C = 10^{-6}$ M.

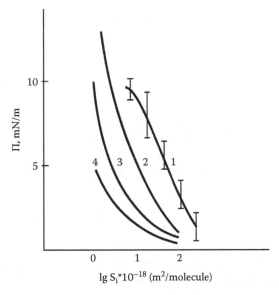

Figure 3.7 $\Pi - \log S_i$ plot of melittin monolayer at salt-free subphase (1) and isotherms calculated by the Davies model (4), the BLP model (3), and the elementary cylinder (2) model.

system. It is easy to show this by writing down the initial expression for calculation of ΔF_1 in the following form:

$$\Delta F_1 = \frac{1}{2} \int_{\infty}^{S_i} \int_{\infty}^{0} \rho \varphi \, dz \, ds, \tag{3.113}$$

and by carrying out integration in indicated limits.

The equation (3.113) is identical to the equation (3.24). Both of these equations should give the same final result. Ahead of integration of the equation (3.113) we shall write out the necessary auxiliary ratios:

$$\rho = -2CN \cdot 1000 \cdot e \cdot sh\frac{e_0\varphi}{kT} \tag{3.114}$$

$$\frac{d\varphi}{dz} = -E = -\frac{2A\sqrt{C}}{\varepsilon_0\varepsilon_1} \cdot sh\frac{e_0\varphi}{2kT} \tag{3.115}$$

$$dz = -\frac{\varepsilon_0\varepsilon_1 d\varphi}{2A\sqrt{C} \cdot sh\dfrac{e_0\varphi}{2kT}}. \tag{3.116}$$

Let's substitute (3.114) and (3.116) to (3.113):

$$\Delta F_1 = \frac{1}{2} \int_{\infty}^{S_i} \int_{0}^{\varphi} \left(\frac{2CN1000e}{2A\sqrt{C}} \frac{sh\dfrac{e_0\varphi}{kT} \varphi \varepsilon_0 \varepsilon_1 d\varphi}{sh\dfrac{e_0\varphi}{2kT}} \right) ds. \tag{3.117}$$

The following expression is equivalent to equation (3.117):

$$\Delta F_1 = \frac{1}{2} \int_{\infty}^{S_i} \int_{0}^{\varphi} \frac{e}{kT} 2A\sqrt{c} \cdot ch\frac{e_0\varphi}{2kT} \, \varphi d\varphi. \tag{3.118}$$

Let's take an intrinsic integral in (3.118) and thus we shall receive:

$$\Delta F_1 = \int_{\infty}^{S_i} \left[\frac{kT}{e} 2A\sqrt{c} \cdot \left(\frac{e\varphi}{2kT} sh\frac{e\varphi}{2kT} - ch\frac{e\varphi}{2kT} \right) \right] ds. \tag{3.119}$$

Let's transform the first term in parentheses in the right part of equation (3.119) according to the following scheme:

$$\frac{e_0\varphi}{2kT} sh\frac{e_0\varphi}{2kT} = \frac{e_0\varphi}{2kT} sh\frac{e}{2kT} \cdot \frac{2kT}{e} Arsh\frac{e_0z}{S_i \cdot 2A\sqrt{C}} = \frac{e_0\varphi}{2kT} \cdot \frac{e_0z}{S_i \cdot 2A\sqrt{C}}, \tag{3.120}$$

and with regard to integration limits we have:

$$\Delta F_1 = \int_{\infty}^{S_i} \left[\frac{e_0 z \varphi}{2 S_i} - \frac{kT}{e} 2A\sqrt{c} \cdot \left(ch\frac{e_0 \varphi}{2kT} - 1 \right) \right] ds. \tag{3.121}$$

Let's compare this result with (3.112) or (3.24) and see that it differs from (3.112) or (3.24) by the factor $1/2 \cdot e_0 z \varphi / S_i$ and by a sign in front of the second term in square brackets. In general, the electrostatic component of surface pressure in ionized monolayer is the derivative with respect to area of reversible compression work of this monolayer, that is:

$$\Pi_{el} = \frac{d\Delta F_1}{dS} = \frac{d}{ds} \int_{\infty}^{S_i} \left[\frac{e_0 z \varphi}{2 \cdot S_i} - \frac{kT}{e} 2A\sqrt{C} \left(ch\frac{e_0 \varphi_s}{2kT} - 1 \right) \right] ds, \tag{3.122}$$

and

$$\frac{d}{ds} \int_{a}^{S} f(s)ds = [f(s)]_s, \tag{3.123}$$

so:

$$\Pi_{el} = \frac{e_0 z \varphi_s}{2 S_i} - \frac{kT}{e} 2A\sqrt{C} \left(ch\frac{e_0 \varphi_s}{2kT} - 1 \right). \tag{3.124}$$

The first term in the right part of equation (3.124) is a «surface» component of specific electrostatic energy in monolayer–subphase system, that is:

$$\frac{1}{2} \frac{e_0 z \varphi_s}{S_i} = \frac{\varepsilon_0 \varepsilon_1}{2} \int_{S} \varphi \left(\frac{d\varphi}{dz} \right) nds. \tag{3.125}$$

The second term in the right part of (3.125) is "volume" component of electrostatic energy, that is:

$$-\frac{2kT}{e} 2A\sqrt{C} \left(ch\frac{e_0 \varphi}{2kT} - 1 \right) = -\frac{\varepsilon_0 \varepsilon_1}{2} \int_{\infty}^{0} E^2 dz = -\int_{0}^{\varphi} \sigma d\varphi, \tag{3.126}$$

where σ – surface density of charges, and φ – electrostatic potential on the interface.

The sum of (3.125) and (3.126) is equal to $\frac{1}{2} \int_{\infty}^{0} \rho \varphi dz$.

Thus, (3.125) is a critical factor. It has fallen out of consideration in Davies model. Accordingly, the electrostatic component of surface pressure appeared smaller than was predicted by the analysis of action of Coulomb forces in the ionized monolayer, considered as a two-dimensional ensemble with discrete structure.

Evaluation of electrostatic component of boundary potential and surface pressure in monolayers with dipole distribution of charges

The dipole-type monolayers consist of non-ionized molecules. This kind of monolayer may be obtained from "neutral" (nondissociated) surface-active substances or from ionized substances under conditions when pH of subphase is still enough to shift proton equilibrium in the polar groups of SAS molecules to non-ionized state.

The Mitchell approach

For the first time the surface pressure analysis in dipole monolayers was carried out by Mitchell [20]. He used a model of rigid rotator, assuming that the monolayer molecules rotate around the axes, perpendicular to water surface, and simultaneously precess in a horizontal plane. The polar groups of SAS monolayer molecules were represented by Mitchell as a continuum of dipoles at the top of the axes. The forces of interaction between molecules were accepted as radial and considered as the functions of four Eiler angles and distance a between the centers of gravity of the molecules. Applying the virial theorem for two-dimensional ensemble of chemically identical molecules with weight m, Mitchell has received the following equation:

$$\Pi\Omega = NkT + \frac{\pi N^2 kT}{\Omega} \int_0^\infty \left[1 - \exp\left(-\frac{U}{kT}\right)\right] a \, da, \qquad (3.127)$$

where Π – surface pressure, Ω – monolayer area, N – Avogadro's number, k – Boltsman constant, T – absolute temperature, U – energy of interaction of each pair of monolayer molecules disposed at a distance a from each other. To analyze the compression isotherms of monolayer using (3.127), the function of U upon packing density of amphiphylic molecules must be known [20]. Mitchell has assumed that all dipoles in monolayer are perpendicularly directed to the interface, and he has found U as a sum of three components:

$$U = U_{rep} + \frac{\mu^2}{\varepsilon_0 \varepsilon_2 a^2} - \frac{\beta}{a^6}. \qquad (3.128)$$

The first term of (3.128) designates the energy of molecules' repulsion at their limit approach. U_{rep} is a disconnected function of a. It tends to infinity if a is less than the diameter of molecules' cross-section, and it tends to zero if a exceeds this size. The second term designates the energy of dipole interaction, where μ is the dipole moment, and ε_0, ε_2 are absolute and relative dielectric permeability in monolayer. The third term corresponds to the contribution of dispersion forces; its magnitude depends upon the length of hydrocarbon radicals of monolayer molecules and is determined by constant β.

To compare the theoretical results with experimental data, it is necessary to know the magnitudes of μ and ε_2. But they are inaccessible to direct measurements. Mitchell has proposed to calculate μ and ε_2 on the basis of the data of dipole moments of amphiphilic molecules measured in vacuum and the data of measurement of boundary potential of real monolayer based on equation:

$$\varphi_b = \varphi_b^0 - \frac{4\pi n \mu_d}{\varepsilon_0 \varepsilon_2},$$

(3.129)

where φ_b – boundary potential jump at known packing density of monolayer molecules, φ_b^0 – boundary potential jump in indefinitely expanded monolayer, μ_d – dipole moment of monolayer molecules measured in vacuum, and n – surface concentration of dipoles inside the monolayer.

Accordingly, the effective dipole moment was determined from the relation:

$$\mu = \mu_d + \frac{\varepsilon_0 \varepsilon_2}{4\pi n} \varphi_b^0.$$

(3.130)

And as a result the following equation for surface pressure was obtained:

$$\Pi = \frac{kT}{S_i} \left\{ 1 + \frac{\pi a^2}{2S_i} \left[1 + \frac{6}{7kT} \left(\frac{2\mu^2}{\varepsilon_0 \varepsilon_2 a^3} - \frac{B}{4a^6} \right) \right] \right\},$$

(3.131)

where S_i – specific area in monolayer, a – the distance between the gravity centers of monolayer molecules.

The first item in the right part of (3.131) is the kinetic component, and the second one describes the sum of dipole and dispersion components. Taking into account the actual dipole constituent, it is easy to be convinced that it is described by the equation:

$$\Pi_{dip} = \frac{6\pi \mu^2}{7\varepsilon_0 \varepsilon_2 S_i^2 a} = \frac{3\mu^2}{8\pi \varepsilon_0 \varepsilon_2 r_0^5}.$$

(3.132)

As well as the values of φ_b^0 in equations (3.129) and (3.130) are inaccessible to direct definition, it is reasonable to analyze the surface pressure in dipole monolayer, proceeding from more simple assumptions.

The models of flat capacitor and cut-out disk

We can accept that effective dipole moment does not vary in monolayer molecules under its compression or expansion. Validity of this assumption is confirmed experimentally for some amphiphilic substances within a wide range of monolayers' packing density [22,23]. In this case the effective dipoles of monolayer molecules can be represented as pairs of opposite charges disposed at a distance δ on axes perpendicular to the interface. Subsequently, the electrostatic component of surface pressure can be found by examining monolayers with different packing density. For example, in the case of condensed monolayers (this situation is represented in Figure 3.8a), the energy of dipole charges distribution can be found as free energy of the flat charged capacitor, and it will be described by the following equation:

$$F_{dip} = -\frac{\varepsilon_0\varepsilon_2}{2}\int_0^v E^2 dV = \frac{e_0^2\delta}{\varepsilon_0\varepsilon_2 S_i}. \tag{3.133}$$

Accordingly, in this case the surface pressure is determined by relation:

$$\Pi_{dip} = -\frac{dF}{dS} = \frac{e_0^2\delta}{2\varepsilon_0\varepsilon_2 S_i^2}. \tag{3.134}$$

In the extremely expanded monolayers (gaseous state) the surface pressure can be found, examining the interaction of each central dipole with infinite numbers of surrounding dipoles (see Figure 3.8b). The energy of interaction of two elementary dipoles is determined by equation:

$$F_i = \mu_\perp E_z = e_0 \delta F_z, \tag{3.135}$$

where

$$E_z = \frac{P}{4\pi\varepsilon_0\varepsilon_2 r_0^3} = \frac{e_0\delta}{4\pi\varepsilon_0\varepsilon_2 r_0^3}. \tag{3.136}$$

Then the specific electrostatic energy F and two-dimensional (surface) pressure in dipole system will be described by the following formulas:

$$F_{dip} = \frac{2\pi p^2}{8\pi\varepsilon_0\varepsilon_2 S_i}\int_{r_0}^\infty \frac{r dr}{r^3} = \frac{e_0^2\delta^2}{4\varepsilon_0\varepsilon_2 S_i r_0} = \frac{\sqrt{\pi}e_0^2\delta^2}{4\varepsilon_0\varepsilon_2 S_i^{3/2}} \tag{3.137}$$

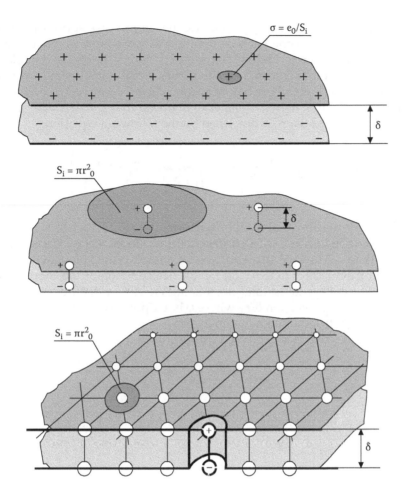

Figure 3.8 (top) The plate capacity model. (middle) The elementary dipole model. (bottom) Cut-out disk model for dipole charge distribution.

$$\Pi_{dip} = \frac{dF}{dS} = \frac{3\sqrt{\pi}e_0^2\delta^2}{8\varepsilon_0\varepsilon_2 S_i^{5/2}} = \frac{3e_0^2\delta^2}{8\pi^2\varepsilon_0\varepsilon_2 r_0^5}. \tag{3.138}$$

But if the value of monolayer packing density is intermediate between the close-packed state and extremely extended state, it is necessary to take into account the discrete distribution of charges (see Figure 3.8c). Thus, the expression for specific energy of the monolayer will have a kind:

$$F = \frac{e_0^2}{4\pi\varepsilon_0\varepsilon_2\delta} - \frac{e_0^2}{2\pi\varepsilon_0\varepsilon_2}\left(\frac{\sqrt{r_0^2 + \delta^2}}{r_0^2} - \frac{1}{r_0}\right). \tag{3.139}$$

This equation was obtained by applying the cut-out disk approach for the lattice of hexagonal type packed dipoles. We have assumed that central dipole is really discrete, and the charges of the other dipoles in monolayer space were smoothly smeared on the planes located at a distance r_0 from central dipole (r_0 is equal to a half of average distance between the dipoles), and δ is the distance between planes of opposite charges. Accordingly, the value of surface pressure is described by the following equation:

$$\Pi_{dip} = \frac{e_0^2}{4\pi\varepsilon_0\varepsilon_2 r_0 S_i}\left(-1 - \frac{1}{\sqrt{1+\delta^2/r_0^2}} + 2\sqrt{1+\delta^2/r_0^2}\right). \tag{3.140}$$

It is easy to show that (3.140) transforms into (3.134) when r_0 is essentially greater than δ, and (3.140) transforms into (3.138) when δ is essentially greater than r_0. Toward this end we can compare the calculated results obtained from (3.134), (3.138), and (3.140) as a function of the ratio of free electrostatic energy to the total electrostatic energy, or in other words, as a function of $\Pi S_{dip}/e_0\varphi$ upon δ/r_0, where φ is a dipole potential generated by dipole charge distribution in monolayers.

In all cases that φ is expressed uniformly, that is,

$$\varphi = \frac{e_0\delta}{\varepsilon_0\varepsilon_2 S_i} = \frac{e_0\delta}{\pi\varepsilon_0\varepsilon_2 r_0^2}, \tag{3.141}$$

we shall have the following expressions for function $\Pi_{dip}S_i/e_0\varphi$:

A. $\dfrac{\Pi_{dip}S_i}{e_0\varphi} = \dfrac{1}{2}$ (3.142)

B. $\dfrac{\Pi_{dip}S_i}{e_0\varphi} = \dfrac{3\delta}{8r_0}$ (3.143)

C. $\dfrac{\Pi_{dip}S_i}{e_0\varphi} = \dfrac{1}{4}\cdot\dfrac{r_0}{\delta}\left(-1 - \dfrac{1}{\sqrt{1+\delta^2/r_0^2}} + 2\sqrt{1+\delta^2/r_0^2}\right)$ (3.144)

     σ – surface charge density
     δ – distance between the charges of opposite signs
     $S_i = \pi r_0^2$ – area per dipole molecule in space of monolayer
Validity of model: $r_0 \ll \delta$.
     S_i – area per dipole in the space of monolayer
     δ – dipole distance
Validity of model: $r_0 \gg \delta$

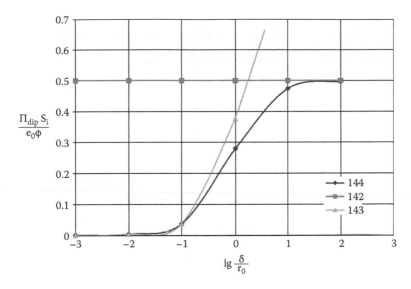

Figure 3.9 $\Pi_{dip}S_i/e_0\varphi$ as a function of $\lg\delta/r_0$ obtained in accord with equations (3.142) to (3.144).

Setting up different magnitudes of δ/r_0, postponing them along axis of abscissas, and calculating magnitudes $\Pi_{dip}S_i/e_0\varphi$ along axis of ordinates, we receive the curves represented on Figure 3.9. As shown, the model of the plane capacitor (equation 3.134) and model of elementary dipoles (equation 3.138) are the special cases of the model described by equation (3.140). At the same time, the calculated values of surface pressure by means of equation (3.140) are in accordance with equation (3.108), which was obtained on the base of virial theorem and cut-out disk approach, that is:

$$\Pi_{dip} = \frac{q^2}{4\varepsilon_0\varepsilon_2 S_i^2}\left(\sqrt{r^2 + \delta^2} + \frac{\delta^2}{\sqrt{r^2 + \delta^2}} - r_0\right). \tag{3.145}$$

Thus, the expression (3.144) is the most suitable for analysis of surface pressure in monolayers with dipole charge distribution. And the boundary potential values in these monolayers can be analyzed on the basis of equation (3.141).

Validity of model: a wide range of r_0 and δ (see Box 3.1).

$$\frac{\Pi_{dip}S_i}{e_0\varphi}$$

$$\lg\frac{\delta}{r_0}$$

BOX 3.1 THE RANGE OF MODELS VALIDITY

PARALLEL PLATE CONDENSER MODEL

$$\varphi_{dip} = \frac{e_0 \delta}{\varepsilon_0 \varepsilon_2 S}$$

$$\Pi_{dip} = \frac{e_0^2 \delta}{2\pi^2 \varepsilon_0 \varepsilon_2 r_0^4}$$

This model is valid for monolayer state analysis at $\delta/r_0 \geq 100$.

ELEMENTARY DIPOLE MODEL

$$\varphi_{dip} = \frac{e_0 \delta}{\varepsilon_0 \varepsilon_2 S}$$

$$\Pi_{dip} = \frac{3}{8} \cdot \frac{e_0^2 \delta^2}{\pi^2 \varepsilon_0 \varepsilon_2 r_0^5}$$

This model is valid for monolayer state analysis at $\delta/r_0 \geq 1$.

CUT-OUT DISK DIPOLE MODEL

$$\varphi_{dip} = \frac{e_0 \delta}{\varepsilon_0 \varepsilon_2 S}$$

$$\Pi_{dip} = \frac{e_0^2}{2\pi^2 \varepsilon_0 \varepsilon_2 r_0^3} \left(-1 - \frac{1}{\sqrt{1 + \frac{\delta^2}{r_0^2}}} + 2\sqrt{1 + \frac{\delta^2}{r_0^2}} \right)$$

This model is satisfied at all ranges of δ/r_0 magnitudes.

IN GENERAL

$$\varphi_{dip} \approx \frac{1}{r_0^4} \div \frac{1}{r_0^5}.$$

Surface bending effects

In the ionized monolayers there are two main factors that can influence the rise of electrostatic part of surface pressure. The first factor is the repulsion of ionized head polar groups of monolayer molecules through the space of the double electric layer of subphase. This part of surface pressure one can calculate on the basis of Davies, Donnan, BLP and other suitable approaches [2,3,8,13]. The second factor may arise from the local bending of subphase area in the space around each ionized head polar group of monolayer-forming molecules (see Figure 3.10). According to this phenomena, the additional number of water molecules will be delivered to air–water interface. So, the additional constituent of surface pressure will appear in the monolayer.

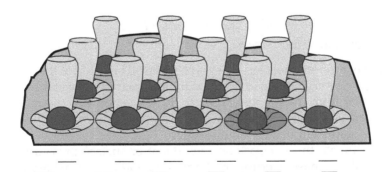

Figure 3.10 Schematic representation of flat and bending space of subphase around the neutral and charged monolayer molecules.

Description of phenomena

The effect of local bending of subphase can play an important part in studying the behavior of multicharged and ionophor monolayers. The main reason for the effect's appearance is the initial electrostatic asymmetry of the air–water system. Proceeding from this circumstance the charges of monolayer-forming molecules are interacted with each other, with its images in air space, and with ions in subphase. Images give rise to additional force acting in the system. This force we can consider as applied to monolayer molecules from the mass centers of image particles and directed to the subphase. As a result the surface-active molecules were shifted to the subphase at a certain distance due to local surface bending. This effect should be taken into account at monolayers state analysis.

For the first time the influence of liquid curvature on its surface tension was studied by Tolman [24]. Analyzing the free energy in systems with flat and curved surfaces, he has received the following relation for surface tension in spherical microscopic drops:

$$\gamma = \gamma_0 \left(\frac{1}{1 + 2\delta/R} \right), \tag{3.146}$$

where γ_0 – surface tension of flat water surface, γ – surface tension of water drop, δ – constant equal to $(0,3 \div 3,0)\ 10^{-10}$ m, and R – radius of a drop.

From (3.146) it follows that the difference in surface tension of flat and spherical water areas can be significant at the comparable values of R and δ.

The local bending of subphase area around the ionized monolayer species

In the water–monolayer–air system the local bending of the subphase area should be accompanied by an additional quantity of water molecules penetrating from the subphase bulk to the space between the monolayer molecules. Estimating the excess of work, one can determine the surface pressure increment in the studied system.

To estimate the electrostatic component of surface pressure arising from local subphase bending in charged monolayers, let us assume initially that monolayer in hypothetical initial state has not exhibited charges. Thus, the surface area in each specific unit of monolayer space is not deformed (Figure 3.11a). It represents a flat circle with radius r_0, which is equal to half the distance between the monolayer-forming molecules. The specific free energy of subphase in this system can be found as a result of

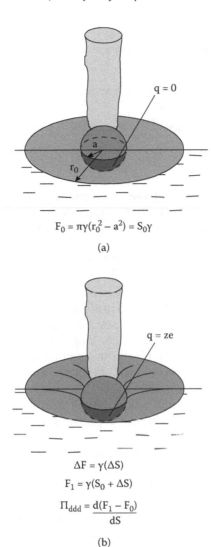

$$F_0 = \pi\gamma(r_0^2 - a^2) = S_0\gamma$$

(a)

$$\Delta F = \gamma(\Delta S)$$
$$F_1 = \gamma(S_0 + \Delta S)$$
$$\Pi_{ddd} = \frac{d(F_1 - F_0)}{dS}$$

(b)

Figure 3.11 Schematic representation of borders around neutral and charged molecules in monolayer at air–water interface.

multiplying surface tension of subphase (γ) by the value of free area (S_0) in the space of monolayer, that is,

$$F_0 = \gamma\pi\left(r_0^2 - a^2\right) = \gamma S_0, \tag{3.147}$$

where a – radius of a head polar group of monolayer-forming molecule.

If the monolayer molecules are ionized, for the above reason it is reasonable to expect some shifting of these molecules to the space of subphase (Figure 3.11b). In this case the area of subphase in each specific unit of monolayer space will be increased by the value of ΔS. And the free energy increment will be:

$$\Delta F = \gamma \Delta S. \tag{3.148}$$

Hence, the total free energy of the system will reach the value:

$$F_1 = \gamma \left(S_0 + \Delta S \right). \tag{3.149}$$

Knowing a difference between free energies before the beginning of deformation of the subphase border and till its end, it is not difficult to find an additional part of surface pressure according to the formula:

$$\Pi_{add} = \frac{d \left(F_1 - F_0 \right)}{dS}. \tag{3.150}$$

In order to determine the magnitude of Π_{add}, it is necessary to know the function of F_1 upon the charge per monolayer-forming molecule, the dimension of these molecules, their packing density at interface, and ionic concentration of subphase. To reach the function of interest it is necessary to solve Poisson–Boltzmann equation for electrostatic field distribution in the space of elementary unit of monolayer–subphase system. The basis of this unit is the specific area per monolayer-forming molecule in the plane of monolayer. The border inside of subphase is the wall of attended cylinder of infinite length.

The cut-out disk model and elementary cylinder model are usually used for solving this task. However, without making a big mistake we can evaluate Π_{add} on the basis of much more simple assumptions like the following:

1. Head polar group of ionized surface-active molecule can be represented as a semi-spherical (Figure 3.12) or spherical (Figure 3.13) arrangement.
2. Location of the depicted arrangement at the interface and its image in air can be represented as shown on the upper parts of Figure 3.12 and Figure 3.13.
3. Taking into account the only effect that is related to electrostatic field action along the axis z, we can calculate the value of force f that is responsible for subphase bending. The following equation will be suitable:

$$f = \frac{\varepsilon_0 \varepsilon_1}{2} \int\limits_0^A E^2 \cos\theta dA, \tag{3.151}$$

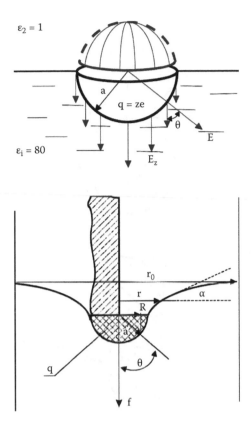

Figure 3.12 Schematic representation of semi-spherical model of ionized head polar group on monolayer molecule.

where $\varepsilon_0\varepsilon_1$ – the dielectric permeability of water media, E – electrostatic field intensity at surface of ionized head polar groups produced by charge images, θ – angle between axis z and any radius–vector outgoing from the center of ionized head polar group to subphase, $A = \pi R^2$ – the local cross-section of head polar group that is expected to be a spherical (or semispherical) arrangement.

Writing down the analytical expressions for E and $\cos\theta$ (the case is shown in Figure 3.12) we shall get:

$$E = \frac{q}{2\pi\varepsilon_0\varepsilon_1 a^2},$$

(3.152)

$$\cos\theta = \frac{\sqrt{a^2 - R^2}}{a},$$

(3.153)

where a – radius of ionized head polar group.

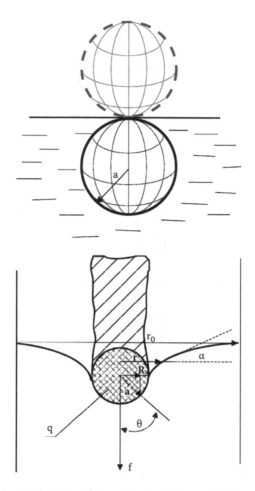

Figure 3.13 Schematic representation of spherical model of ionized head polar group.

After integrating of (3.151) and taking into account (3.152) and (3.153) we shall receive:

$$f = \frac{1}{4\pi\varepsilon_0\varepsilon_1} \cdot \frac{q^2}{a^5} \int_0^a \sqrt{a^2 - R^2} \cdot R\,dR = \frac{q^2}{12\pi\varepsilon_0\varepsilon_1 a^2}. \tag{3.154}$$

Under elastic deformation of water surface the angle α (Figure 3.12 and Figure 3.13) and the force f can be represented by the ratio:

$$\sin\alpha = \frac{f}{2\pi r\gamma}, \tag{3.155}$$

where r – distance through the normal from axis z up to any point of a surface bending in each elementary unit of monolayer. Accordingly, the area increment due to subphase bending will be:

$$dS = \frac{2\pi r dr}{\cos\alpha}.\tag{3.156}$$

Expressing $\cos\alpha$ through the values of force–vector and radius r, we shall get:

$$dS = \frac{2\pi r dr}{\sqrt{1 - \left(\dfrac{f}{2\pi r \gamma}\right)^2}}\tag{3.157}$$

or:

$$S = \int_a^{r_0} \frac{2\pi r^2 dr}{\sqrt{r^2 - \left(\dfrac{f}{2\pi \gamma}\right)^2}},\tag{3.158}$$

where S – area of curved surface in elementary unit of monolayer.

As long as the difference of flat and curved surfaces is proportional to the change of free energy of studied system, then subtracting from (3.158) the magnitude of flat area (S_0), and multiplying the obtained result by γ, we shall get:

$$\Delta F = \gamma \left(\int_a^{r_0} \frac{2\pi r^2 dr}{\sqrt{r^2 - \left(\dfrac{f}{2\pi \gamma}\right)^2}} - \int_a^{r_0} 2\pi r dr \right).\tag{3.159}$$

And taking the derivative of ΔF with respect to S, we shall find an additional component of surface pressure arising as a result of subphase bending:

$$\Pi_{add} = \frac{d\Delta F}{dS} = \gamma \left[\frac{r_0}{\sqrt{r_0^2 - \left(\dfrac{f}{2\pi \gamma}\right)^2}} - 1 \right].\tag{3.160}$$

Thus, in ionized monolayer with disposition of ionized molecules as depicted by Figure 3.12 the surface pressure increment caused by the effect of local bending of subphase will be increased proportionally by the number of ionized charges per monolayer molecule and will be decreased proportionally by the radius of ionized head polar groups and distance between these groups, that is,

$$\Pi_{add} = \gamma \left[\frac{r_0}{\sqrt{r_0^2 - \left(\dfrac{(ze)^2}{24\pi^2 \varepsilon_0 \varepsilon_1 a^2 \gamma} \right)^2}} - 1 \right]. \tag{3.161}$$

If the disposition of polar groups will be assumed to be like that shown in Figure 3.13 then the expression for f will get the form:

$$f = \frac{(ze)^2}{16\pi \varepsilon_0 \varepsilon_1 a^2}. \tag{3.162}$$

And the increment of surface pressure will be described by equation:

$$\Pi_{add} = \gamma \left[\frac{r_0}{\sqrt{r_0^2 - \left(\dfrac{(ze)^2}{32\pi^2 \varepsilon_0 \varepsilon_1 a^2 \gamma} \right)^2}} - 1 \right]. \tag{3.163}$$

If the monolayer consists of the charges inside of sequence of ion–peptide complexes (e.g., valinomycin monolayers on subphase with an excess of KCl), the local bending of subphase will take place due to the interaction of surplus of complexone charge with ion charges of subphase. To do the calculation of force f in this case (see Figure 3.14), the following equation may be used:

$$f = \frac{q^2}{16\pi \varepsilon_0 \varepsilon_2 \delta^2}, \tag{3.164}$$

where q – excess of charge per complexone molecule, ε_2 – relative dielectric permeability in space of monolayer ($\varepsilon_2 \ll 80$), and δ – height of excess charge disposition relatively to plane water surface.

Equation (3.164) is valid when $\delta \leq r_0$. If r_0 is comparable for the δ then the force f can be calculated on the basis of the following equation:

$$f = \frac{q^2}{\pi \varepsilon_0 \varepsilon_1 r_0^4} \int_0^{r_0} \left\{ 1 + \sum_{i=1}^{\infty} \exp(-k_i \delta) \frac{I_0(k_i r)}{I_0(k_i r_0)^2} \right\}^2 r \, dr. \tag{3.165}$$

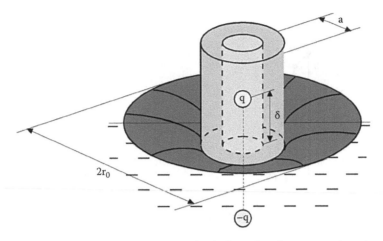

Figure 3.14 Schematic representation of subphase bending around molecule of charged surface-active complexone.

This equation is received as result of electrostatic field distribution analysis in the model of elementary circular cylinder, which simulates an elementary cell in monolayer formed from ion–peptide complexones.

Surface bending effect in monolayers of surface-active complexones

The monolayers of surface-active complexones exhibit the well-defined ability to bind selectively the ions from subphase [25–28]. The cations' binding causes the surface pressure to increase and the boundary potential jump in the monolayer [25,26]. The anions' binding is accompanied by the surface pressure increasing, but in this case the boundary potential jump in monolayer is decreased [29].

The peculiarity of interaction of surface-active complexones with ions of subphase is followed by more powerful generation of boundary potential and surface pressure, than in the case of simple surface-active substances [26,29]. For example, in the case of valinomycin monolayers, the boundary potential and the surface pressure are increased as shown in Figure 3.15 when the concentration of potassium chloride in subphase is raised from zero up to 4.0 M. Evidently the phenomenon is caused by potassium ions penetrating from subphase into cyclodepsypeptide skeletons of valinomycin molecules and by interaction of these ions among themselves and with subphase media through the low-dielectric permeability space (hydrophobic zone of monolayer) ($\varepsilon_2 \ll 80$).

Let's show that surface pressure in monolayer consisting of charged ion–peptide complexes is essentially determined by local bending of subphase surface. With this purpose the total value of surface pressure of monolayer

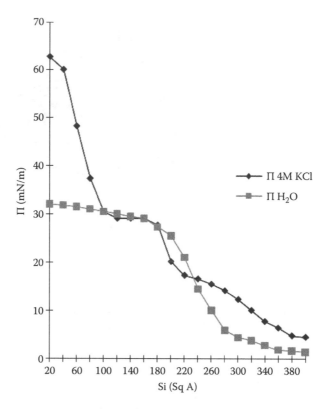

Figure 3.15 Surface pressure Π and boundary potential φ_b versus area per molecule in valinomycin monolayers on H_2O and 4M KCl subphases.

will be represented as the sum of four constituents: kinetic (Π_{kin}), dispersion (Π_{disp}), lateral electrostatic (dipole) (Π_{dip}), and normal electrostatic (Π_{add}), that is, caused by surface bending of subphase in each elementary cell of monolayer.

$$\Pi = \Pi_{kin} + \Pi_{disp} + \Pi_{dip} + \Pi_{add} \qquad (3.166)$$

The magnitudes of the first three components in the right part of (3.166) can be calculated on the basis of equations (3.3), (3.4), and (3.140). The Π_{add} component, which is arised from local bending of subphase surface, can be calculated on the basis of equation (3.164) or (3.165), if the function of f upon distance r_0 between ion–peptide complexes is known.

To find the relationship between f and r_0 in elementary unit of monolayer space, we can represent this unit as the circular cylinder (see Figure 3.16). The base of the cylinder will be the interface between the surface and air, and the element of the cylinder will be a geometrical place

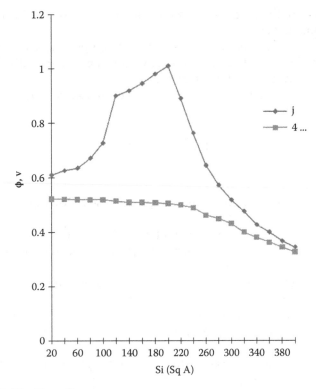

Figure 3.15 (Continued)

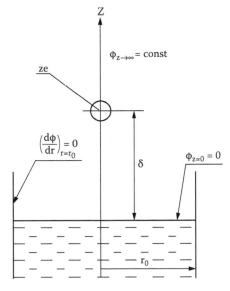

Figure 3.16 Elementary cylinder model.

of points where the potential gradients of adjacent ion–peptide complexes become equal to zero.

The cation penetrated into the space of compexone skeletons is represented in Figure 3.16 as a sphere in which the center is displaced at a distance δ from the subphase surface.

For calculations it is necessary to solve the Laplace equation for the circular cylinder:

$$\frac{d^2\varphi}{dr^2} + \frac{1 \cdot d\varphi}{r \cdot dr} + \frac{d^2\varphi}{dz^2} = 0. \tag{3.167}$$

The boundary conditions in this case are the following:

1. The radial component of potential gradient ($d\varphi/dr$) is equal to zero at $r=r_0$; (because of symmetric packing of monolayer-forming molecules).
2. The electric potential on the base of cylinder is equal to zero (interface area is expected to be an equipotential).
3. The electric potential on infinity in air is constant value (all forces lines of a field strength produced by monolayer charges are absorbed by subphase).
4. The current flow of electrostatic field at subphase area is equal to the value of ion charge in the space inside complexone molecule, that is,

$$2\pi\varepsilon_0\varepsilon_2 \int_0^{r_0} \left(\frac{d\varphi}{dz}\right)_{z=0} rdr = q. \tag{3.168}$$

Partial integral of equation (3.167) is reasonable to search as a function:

$$\varphi = R \cdot U, \tag{3.169}$$

where R and U depend upon their coordinates only, that is, r and z:

$$\frac{d^2\varphi}{dr^2} = U\frac{d^2R}{dr^2}$$

$$\frac{d\varphi}{dr} = U\frac{dR}{dr} \tag{3.170}$$

$$\frac{d^2\varphi}{dz^2} = R\frac{d^2U}{dz^2}$$

Due to linearity the equation (3.167) should be satisfied by the sum of Laplace products (3.169), where the random coefficient model is satisfied for boundary conditions.

Upon substituting (3.170) into (3.167) and dividing all items of (3.167) into (3.169), the following equation will be obtained:

$$\frac{I \cdot d^2R}{R \cdot dr^2} + \frac{1 \cdot dR}{Rr \cdot dr} + \frac{I \cdot d^2U}{U \cdot dz^2} = 0. \tag{3.171}$$

It is reasonable to rewrite the ratio (3.171) as the system of two equations where each of them depends upon their own variables, that is:

$$\frac{I \cdot d^2U}{U \cdot dz^2} = K^2, \tag{3.172}$$

$$\frac{d^2R}{dr^2} + \frac{I \cdot dR}{r \cdot dr} + K^2 = 0. \tag{3.173}$$

This system provides the partial solutions for U and R. Within the range of z from $z = 0$ until $z = \delta$, the equation (3.172) has the following partial solutions:

$$U - B \cdot z, \tag{3.174}$$

$$U = Cshkz + Dchkz, \tag{3.175}$$

where B, C, D – the constants.

When $z = 0$, U is equal to zero by definition, and accordingly, $D = 0$. As a result for $z = \delta$, we shall get:

$$U_\delta = B\delta, \tag{3.176}$$

$$U = Cshk\delta. \tag{3.177}$$

Within the range of $z = \delta$ to $z \to \infty$ we shall get:

$$U = U_\delta \exp\left[-k(z - \delta)\right]. \tag{3.178}$$

Substituting the expression for U_δ from (3.176) into (3.178), we may put together the following solutions:

$$U = C \cdot shk\delta \cdot \exp[-k(z-\delta)] \qquad (3.179)$$

or

$$U = \frac{1}{2}C \cdot [\exp(2k\delta)-1] \cdot \exp(-kz). \qquad (3.180)$$

The general solution for R has a form of the Bessel function of first genus and zero order [30]:

$$R = \sum_{0}^{\infty} A_i I_0(k_i r_0), \qquad (3.181)$$

where $k_i = \mu_i/r_0$, μ_i – roots of the Bessel functions $I_1(x)$, A_i – coefficient which takes into account a preservation of the condition of electrostatic field strength intensity from a charge produced by complexone.

The sum of solution (3.181) regards R and U as a searching solution of equation (3.171). So, when $z < \delta$ the potential value is described by the expression:

$$\varphi = BA_0 z I_0(k_0, r) + C\sum_{1}^{\infty} A_i shk_i z I_0(k_i, r), \qquad (3.182)$$

and when $z > \delta$ – by expression

$$\varphi = BA_0 \delta I_0(k_0, r) + \frac{1}{2}C[\exp(2k_i\delta)-1] \cdot \sum_{1}^{\infty} A_i \exp k_i z I_0(k_i, r). \qquad (3.183)$$

As long as $I_0(0) = 1$ by definition, the equations (3.182) and (3.183) can be simplified as follows:

$$\varphi = BA_0 z + C\sum_{1}^{\infty} A_i shk_i z I_0(k_i, r) \qquad (3.184)$$

$$\varphi = BA_0 \delta + \frac{1}{2}C[\exp(2k_i\delta)-1] \cdot \sum_{1}^{\infty} A_i \exp k_i z I_0(k_i, r) \qquad (3.185)$$

In (3.184) and (3.185) we shall find A_0 and A_i. According to this purpose we shall write down the expressions for potential gradient adjoining to $z = \delta$:

$$\left(\frac{d\varphi}{dz}\right)_{z<\delta} = BA_0 + C \cdot \sum_1^{\infty} A_i k_i chk_i z I_0(k_i,r).$$
(3.186)

$$\left(\frac{d\varphi}{dz}\right)_{z>\delta} = -\frac{1}{2} C \cdot [\exp(2k_i\delta) - 1] \cdot \sum_1^{\infty} A_i k_i \exp(-k_i z) I_0(k_i,r).$$
(3.187)

Difference of these gradients is equal to potential gradient those is expressed by generalized delta function:

$$BA_0 + C \sum_1^{\infty} A_i k_i \exp(k_i\delta) I_0(k_i,r) = \frac{q\delta_{(r)}}{\varepsilon_0\varepsilon_2},$$
(3.188)

where q – charge value, $\varepsilon_0\varepsilon_2$ – dielectric constant, and $\delta_{(r)}$ – generalized delta function.

Now we can multiply both parts of (3.188) by $I_0(k_ir)rdr$ and integrate the obtained expression. In this case all members $k_i \neq i$ will drop out, and as a result we shall get:

$$\frac{qI_0(0)}{2\pi\varepsilon_0\varepsilon_2} = BA_0 \int_0^{r_0} [I_0(0)]^2 rdr = \frac{1}{2} BA_0 r_0^2.$$
(3.188)

Hence,

$$A = \frac{q}{\pi\varepsilon_0\varepsilon_2 Br_0^2}.$$
(3.190)

Similarly,

$$\frac{qI_0(0)}{2\pi\varepsilon_0\varepsilon_2} = CA_i k_i \exp(k_i\delta) \cdot \int_0^{r_0} [I_0(k_ir)]^2 \cdot rdr,$$
(3.191)

$$\frac{q}{2\pi\varepsilon_0\varepsilon_2} = \frac{CA_i k_i \exp(k_i\delta) r_0^2}{2} \cdot [I_0(k_ir)]^2,$$
(3.192)

$$A_i = \frac{q\exp(-k_i\delta)r_0^2 \cdot 1}{\pi\varepsilon_0\varepsilon_2 r_0^2 Ck_i \cdot \left[I_0(k_i r)\right]^2},$$ (3.193)

Substituting the expressions for coefficients A_0 and A_i into equations (3.184) and (3.185), we shall receive:

$$\varphi_{z<\delta} = \frac{q}{\pi\varepsilon_0\varepsilon_2 r_0^2} \cdot \left[z + \sum_0^\infty \frac{\exp(k_i\delta)}{k_i} shk_iz \frac{I_0(k_i,r)}{\left[I_0(k_i,r)\right]^2}\right]$$ (3.194)

$$\varphi_{z>\delta} = \frac{q}{\pi\varepsilon_0\varepsilon_2 r_0^2} \cdot \left[\delta + \sum_0^\infty \frac{sh(-k_i\delta)}{k_i}\exp(-k_iz)\frac{I_0(k_i,r)}{\left[I_0(k_i,r)\right]^2}\right].$$ (3.195)

Thus we have got a solution of the equation (3.167) in terms of $\varphi - \delta$. Taking the derivative of (3.194) with respect to z we shall receive:

$$\left(\frac{d\varphi}{dz}\right)_{z=\delta} = \frac{q}{\pi\varepsilon_0\varepsilon_2 r_0^2} \cdot \left[1 + \sum_0^\infty \exp(-k_i\delta)\frac{I_0(k_i,r)}{\left[I_0(k_i,r_0)\right]^2}\right].$$ (3.196)

In such a way the magnitude of attractive force f between charge and subphase arisen from action of this charge in the space of hydrophobic media should be described by

$$f = \frac{\varepsilon_0\varepsilon_2}{2}\int_0^{S_0}\left(\frac{d\varphi}{dz}\right)_{z=0}^2 dS$$ (3.197)

or

$$f = \frac{2\pi\varepsilon_0\varepsilon_2}{2}\int_0^{r_0}\left(\frac{d\varphi}{dz}\right)_{z=0}^2 rdr.$$ (3.198)

Finally, we shall get:

$$f = \frac{q}{\pi\varepsilon_0\varepsilon_2 r_0^4}\int_0^{r_0}\left\{1 + \sum_{i=1}^\infty \exp(-k_i\delta)\frac{I_0(k_ir)}{\left[I_0(k_ir_0)\right]^2}\right\} rdr.$$ (3.199)

The analysis of this equation shows that at $\delta \ll r_0$, the equation (3.199) transforms into the following:

$$f = \frac{q^2}{16\pi\varepsilon_0\varepsilon_2\delta^2}.$$ (3.200)

Making a substitution of (3.199) or (3.200) into (3.160) we can get the magnitudes of Π_{add} in the case of action of charged ion–peptide complexones in the space of monolayer, that is,

$$\Pi_{add} = \frac{d\Delta F}{dS} = \frac{d\cdot\gamma}{2\pi r dr_0} \cdot \left[\int_a^{r_0} \frac{2\pi r^2 dr}{\sqrt{r^2 - \left(\frac{f}{2\pi\gamma}\right)^2}} - \int_a^{r_0} 2\pi r dr\right] = \gamma \cdot \left[\frac{r_0}{\sqrt{r_0^2 - \left(\frac{f}{2\pi\gamma}\right)^2}} - 1\right].$$

(3.201)

Comparison of calculated results and experimental data on valinomycin monolayers

Valinomycin has no ionized polar groups in its molecules. But in monolayers formed from this substance on the surface of concentrated KCl solution, valinomycin molecules exhibit the net charge behavior. This phenomenon is due to potassium cations penetrating into cyclodepsipeptide skeleton. The arisen complex is shown schematically in Figure 3.17. The equivalent opposite charges (anions) are located in subphase. The distance between depicted charges is equal to "δ". In such a way the "chemically" induced dipoles are the reason for net charge behavior of this object.

One is able to analyze the surface pressure in stated monolayer as the sum of kinetic, dispersion, and dipole constituents, including surface-bending additive. One is able to calculate the kinetic constituent of surface pressure by equation (3.3), and the dispersion constituent by equation (3.4). To calculate the "dipole" additive one needs to know the "dipole" distance inside the monolayer species. To get this knowledge the follow equation can be used:

$$\varphi_b = \varphi_{dip} = \frac{e_0\delta}{\varepsilon_0\varepsilon_2 S_i} = \frac{e_0\delta}{\varepsilon_0\varepsilon_2\pi r_0^2}.$$ (3.202)

All symbols in this equation have their ordinary meanings. Equation (3.202) allows finding the ratio δ/ε_2. when the magnitudes of φ_b and S_i are known. For merely expanded valinomycin, monolayer φ_b is proportional to S_i, and required magnitudes of these items can be selected from $\varphi_b - S_i$ plots depicted in Figure 3.15.

Assuming that relative dielectric permeability in valinomycin monolayer is equal to 4 (i.e., it is the typical value of ε_2 at water–air interface

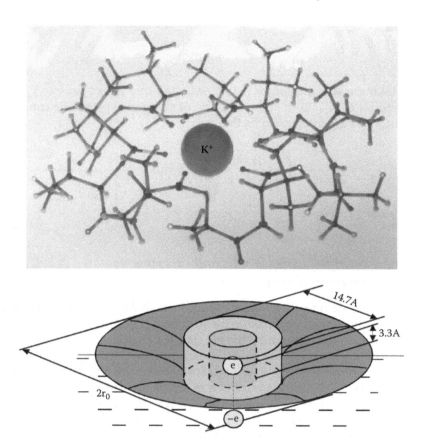

Figure 3.17 Molecule model of valinomycin and schematic representation of elementary cell in its monolayer.

covered by amphiphilic monolayer) and substituting the selected values of φ_b and S_i into (3.202), one will get:

$$\delta = \frac{\varphi_{dip}\varepsilon_0 S_i}{e_0} = 3,3 \cdot 10^{-10}(m). \tag{3.203}$$

The magnitude of $\delta = 3,3 \cdot 10^{-10}\,m$ is lower than the height of valinomycin molecule. But this magnitude is still enough to initiate the local bending effect around each complexone molecule in merely compressed monolayers.

In such a way, by substituting the calculated value of δ into (3.140), one will evaluate the dipole part of surface pressure in complexone monolayer. And substituting δ into (3.200) and (3.160), one will evaluate the additive to the surface pressure arisen from the subphase bending.

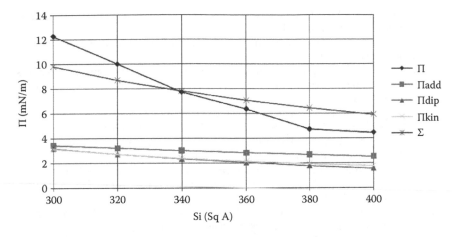

Si	П	Пadd	Пdip	Пkin	Σ
300	12.27	3.4534	3.20927	3.16338	9.826
320	10	3.2228	2.73841	2.7416	8.7028
340	7.72	3.0211	2.35889	2.41906	7.7991
360	6.36	2.8432	2.04914	2.16442	7.0568
380	4.72	2.6851	1.79349	1.95829	6.4369
400	4.45	2.5437	1.58036	1.788	5.912

Figure 3.18 П versus Si valinomycin monolayers on H_2O and 4M KCl and calculated values of П for charged state of this monolayer at Si = 400 ÷ 280 A2.

The calculated plots of the sum of Π_{kin}, Π_{disp}, Π_{dip}, Π_{add} versus S_i for monolayer of valinomycin formed on 4M KCl as subphase are represented in Figure 3.18. Comparison of calculated data with the experimental isotherm shows the satisfactory agreement when all factors responsible for surface pressure arising in monolayer have been taken into account.

References

1. Jaycock I., Parfitt G.D. *Chemistry of Interfaces.* New York: Hasted Press, 1981.
2. Adamson A., Fizicheskaya Khimiya Poverkhnostey. *Physical Chemistry of Surfaces.* Moscow: Mir, 1979. (Russian translation).
3. Matijevic E., Pethica B.S. The properties of ionized monolayers. *Trans. Faraday Soc.* 1958. 54:1382–1407.
4. Gaines G.L., Jr. From monolayer to multilayer: Some unanswered questions. *Thin Solid Films.* 1980. 68:1–5.
5. Gaines G.L., Jr. Insoluble monolayers at liquid–gas interfaces. New York: Interscience, 1966.
6. Shinoda K., Nakagava T., Tamamusi B., Isemura T. *Kolloidnie poverkhnostnie veschestva.* Edited by A.B. Taubman, Z.N. Markina. Moscow: Mir, 1966. (Russian translation).

7. Moelwin-Hughes E.A. *Physical Chemistry.* Edited by Y.I. Gerasiomov. Moscow: Inostrannaya Literatura, 1962. (In Russian).
8. Davies J.T. A surface equation of state for charged monolayers. *J. Coll. Sci.* 1956. 11(4):377–390.
9. Davies J.T. The application of the Gibbs equation to charged monolayers, and their desorption from the oil-water interface. *Trans. Faraday Soc.* 1952. 48:1052–1061.
10. Phillips J.N., Rideal E. The influence of electrolytes on gaseous monolayers. I. Neutral films. *Proc. Roy. Soc., Ser. A.* 1955. 232(1189):149–158.
11. Phillips J.N., Rideal E. The influence of electrolytes on gaseous monolayers. II. Ionized films. *Proc. Roy. Soc.* 1955. 232(1189):159–172.
12. Standish M.M., Pethica B.A. Surface pressure and surface potential study of a synthetic phospholipid at air/water interface. *Trans. Far. Soc.* 1968. 64(544):1113–1122.
13. Levin S., Robinson K. Ion-size statistics and the Esin-Markov coefficient. *Electroanalytical Chem. and Interfac. Chem.* 1975. 58:19–29.
14. Bell G.M., Levine S. Surface free energy and surface pressure of electrolytes and charged monolayers. *Z. Phys. Chem.* 1966. 231(5/6):289–309.
15. Bell G.M., Levine S., Pethica B.A., Stephens C. The surface pressure of ionized monolayers; a reply to Payens. *J. Coll. Inter. Sci.* 1970. 33(3):482–483.
16. Levine S., Mingins J., Bell G.M. The discrete-ion effect in ionic double-layer theory. *J. Electroenel. Chem.* 1967. 13:280–329.
17. Levine S., Bell G.M., Calvert D. The discreteness-of-charge eggect in electric double layer theory. *Canad. J. Chem.* 1962. 40:518–538.
18. Bell G.M., Levine S., Pethica B.A. The surface pressure of ionized monolayers. *Trans. Farad. Soc.* 1962. 58:904–917.
19. Levine S., Mingins J., Bell G.M. The discrete-ion effect and surface potentials of ionized monolayers. *J. Phys. Chem.* 1963. 67(10):2095–2105.
20. Mitchell J.S. Some aspects of the equation of state of monolayers. *Trans. Farad. Soc.* 1935. XXXI(8):980–986.
21. Gevod V.S., Reshetnyak S.I., Gevod S.V. et al. Raschot micropotentsialov i electrostaticheskikh polei v ionizirovannikh amphiphilnikh monosloyakh na osnove chislennogo resheniya uravneniya Puassona-Boltzmana. *Voprosi khimii i khim. tekhnologii.* 1999. 1:79–81. (In Russian).
22. Ksenzhek O.S., Gevod V.S.. On the ionic double layer structure of the BLM film from the data obtained on monolayers. *Extended Abstracts of the 37 Meeting of the International Society of Electrochemistry.* Vol. IV, Vilnius. 1986:452–454.
23. Monolayers. Advances in chemistry series. No. 144. Washington: *American Chem. Soc.* 1975.
24. Tolman R.C. Consideration of the Gibbs theory of surface tension. *J. Chem. Phys.* 1948. 16(8):758–774.
25. Birdi K.S., Gevod V.S. Monolayers of globular proteins and membrane proteins (A: Melittin; B: Valonomycin) a membrane model system. *Proceedings of Biophysics School of Membrane Transport*, Zakopane, Poland, 1984. Zakopane. 1984:3–40.
26. Kemp G., Wenner C.E. Interaction of valonimycin with cations at the air-water interface. *Biochim. Biophys. Acta.* 1972. 282:1–7.

27. Caspers J., Landuyt-Caufries M., Deleers M., Ruysschaert J.M. The effect of surface charge density on valinomycon-K^+ complex formation in model membranes. *Biochim. Biophys. Acta.* 1979. 554:23–38.

28. Shlyakhter T.A., Lev A.A. Ismenenie sostoyaniya sloya molekul litsitina na granitse rasdela geptan – vodnie rastvori pri vvedenii ionophora – valinomitsina. *Tsitologiya.* 1980. XXII(10):1193–1199. (In Russian).

29. Gevod V.S., Birdi K.S. Melittin and the 8-26 fragment: differences in ionophoric properties as measured by monolayer method. *Biophys. J.* 1984. 45:1079–1083.

30. Ango A. *Mathematica dlya electro- i radioinjenerov.* Moscow: Nauka, 1967. (In Russian).

chapter four

Ion-exchange and ion-specific effects in lipid monolayers

V. S. Gevod, I. L. Reshetnyak, and S. V. Gevod

Contents

Introduction

The surface charge density of lipid monolayers is the function of the ionization degree of their molecules. In the case of monolayers formed from fatty acids, the surface charge density increases on alkaline subphases and decreases on acidic subphases. In fatty amine monolayers, the surface charge density has the opposite sign and reaches maximum on the alkaline subphases. In monolayers formed from zwitterionic lipids (which molecules contain both basic and acidic groups), the charge density reaches maximal value at extremely high and extremely low pH. But at neutral pH the monolayers of zwitterionic lipids act as the neutral (nonionized) surface films. The reason for the phenomenon is the mutual compensation of ionized charges inside the monolayer space.

In studying ionized monolayers one needs to know how pH and dissolved salts of subphase affect the ionization degree of monolayer molecules. From this point of view the properties of monolayers formed from stearic acid and zwitterionic phospholipids (dipalmytoilphosphatidylserine and dipalmytoilphosphatidylcholine) were considered as pH-dependent.

The study of double electric layer of stearic acid monolayer and its derivatives at the air–water interface

The majority of cells and tissues contain free fatty acids [1,2]. Usually they serve as the initial substances for the synthesis of fats, glycolipids, ethers of cholesterol, and so forth [1–3]; but they are also able to perform a regulator function [4,5]. The biochemistry of fatty acids is studied in detail [6]. The physical properties of these substances in biological membranes are less investigated. In particular, additional research is required to understand their influence on the stability of lipid matrix, their participation in the processes of lateral protons transfer, and their role in the regulation of the intra-membrane field [2,3,7,8].

Properties of fatty acids in artificial lipid systems were studied first of all using liposomes and monolayers [9–12]. It was observed that their properties depend upon the length and the structure of hydrocarbon radicals in fatty acid molecules and the ionization degree of polar groups of these molecules. The structural factors (i.e., the length and the branching of hydrocarbon radicals, the presence of double bonds) determine the packing density of amphiphilic molecules in bilayers and liposomes. Thus, they have an influence upon their permeability for liquids and gases. The building up of intermolecular hydrogen bonds and charge shielding of polar heads of monolayers molecules by ions of aqueous phase have a certain influence upon the intra-membrane field.

The influence of pH and the concentration of one- and bivalent cations upon dipole and surface components of boundary potential and surface pressure were studied in the condensed monolayers of stearic acid and its derivatives (anilide of stearic acid and 1,5-bis(4-stearoilaminophenyl)-3-phenylformazane).

Basic theory

The boundary potential jump (φ_b) in any amphiphilic monolayer includes dipole (φ_{dip}) and surface (φ_{sur}) constituents. The value of φ_{dip} depends upon the chemical structure of amphiphilic molecules, their conformation, and packing density. The value of φ_{sur} depends upon the quantity of ionized head polar groups in monolayer molecules, their packing density, the sign of their charge, and ionic strength of the subphase. The degree of monolayer ionization shifts over a wide range, depending upon pH of the subphase, and in this way the monolayers are transformed into neutral («dipole») or completely ionized boundary films.

In the close-packed monolayers of dipole type, the value of boundary potential is described by the Helmholtz equation [11]:

$$\varphi_b = \varphi_{dip} = \frac{e\delta}{\varepsilon_0\varepsilon_1 S_i}, \tag{4.1}$$

where e – the elementary charge, δ – the effective dipole length in the direction perpendicular to subphase surface, ε_0 – the absolute dielectric permeability, ε_1 – the dielectric permeability of monolayer space, and S_i – the specific area in monolayer.

In the case of ionized monolayers a double electrical layer is induced additionally in the subphase by surface charge. And it makes a contribution to the boundary potential. The value of φ_{sur} is described by the Guoy–Chapman equation [13]:

$$\varphi_{sur} = \frac{2kT}{e} Arsh\frac{\sigma}{\sqrt{8INkT\varepsilon_0\varepsilon_2}}. \tag{4.2}$$

In such a way the boundary potential in any amphiphilic monolayer can be represented as the sum of φ_{dip} and φ_{sur}, that is:

$$\varphi_b = \frac{e\delta}{\varepsilon_0\varepsilon_1 S_i} + \frac{2kT}{e} Arsh\frac{\sigma}{\sqrt{8INkT\varepsilon_0\varepsilon_2}}, \tag{4.3}$$

where k – the Boltzmann constant, T – the absolute temperature, N – Avogadro's number, ε_2 – the dielectric permeability of the subphase, σ – the surface charge density, I – the ionic strength of the subphase $\left(I = 1/2\sqrt{c_i z_i^2}\right)$, z – the charge of i-kind ions, and c_i – the ions concentration in subphase.

The surface pressure in dipole and ionized monolayers alongside with properly electrostatic constituents includes dispersion and kinetic components. They arise due to the action of the Van der Waals attraction forces and the thermal repulsion forces between the monolayer molecules. So, we can write:

$$\Pi_1 = \Pi_{dip} + \Pi_{kin} + \Pi_{disp}, \tag{4.4}$$

$$\Pi_2 = \Pi_{dip} + \Pi_{sur} + \Pi_{kin} + \Pi_{disp}, \tag{4.5}$$

where Π_1 and Π_2 – ranks among the dipole and ionized monolayers, respectively; Π_{disp} and Π_{kin} – the dispersion and kinetic constituents of surface pressure; and Π_{dip} and Π_{sur} – the electrostatic constituents of surface pressure, which are stipulated by the presence of dipoles and free charges in the monolayer molecules.

The analytical expressions for the components Π, which enter into the right-hand side of the equations (4.4) and (4.5), are as follows:

$$\Pi_{kin} = \frac{kT}{S_i - S_0},\tag{4.6}$$

$$\Pi_{disp} = \frac{-\beta}{d^6},\tag{4.7}$$

$$\Pi_{dip} = \frac{e}{4\pi^2\varepsilon_0\varepsilon_1 r_0^3}\left(2\sqrt{1+\frac{\delta^2}{r_0^2}} - \frac{1}{\sqrt{1+\frac{\delta^2}{r_0^2}}} - 1\right),\tag{4.8}$$

$$\Pi_{sur} = \frac{2kT}{e}\sqrt{8INkT\varepsilon_0\varepsilon_2}\left(chsh^{-1}\frac{\sigma}{\sqrt{8INkT\varepsilon_0\varepsilon_2}} - 1\right),\tag{4.9}$$

where β – the constant of van der Waals interaction of amphiphilic molecules in the monolayer space, d – the distance between these molecules, and S_0 – the cross-section area of amphiphilic molecules. The other symbols have their generally accepted designations.

The equations (4.1)–(4.9) show that the boundary potential and the surface pressure in the ionized monolayer depend upon many factors, such as $\sigma, \delta, \beta, I, S_i, \varepsilon_1, \varepsilon_2, T$. Therefore, the study of such monolayers is a complex problem. However, it becomes essentially simpler if one will examine not absolute values of Π_i and φ_b, but their variation upon $\sum c_i$ at constant T and S_i. In this case Π_{kin} and Π_{disp} are the constants, and $\Delta\Pi$ and φ_b depend only upon the changes of the sum of dipole and "free charges" components Π and φ. They are described by the equations (4.1)–(4.2) and (4.7)–(4.8).

We can also expect that Π_{dip} and φ_{dip} should be the constant values in the monolayers formed from molecules with rigid structure. In this case the subphase should have an influence upon the values Π_{sur} and φ_{sur}. They are connected with σ by the equations (4.2) and (4.9), respectively. But if the monolayers consist of conformation-labile molecules, the function of Π and φ_b upon $\sum c_i$ will be dependent upon the influence of subphase on the parameter δ. In this case the values of Π_{dip} (and φ_{dip}) should be equal to the differences between the experimentally registered values of Π (or φ_b) and the calculated values of Π_{sur} (or φ_{sur}), obtained by equations (4.2) and (4.9).

The equation of ionization of fatty acid molecules, as applied to monolayer, is of the following form:

$$RCOO^- + H^+ \leftrightarrow RCOOH\tag{4.10}$$

Respectively, the expression for the ionization constant of monolayer substance turns out as follows:

$$K_{ion} = \frac{[RCOO^-] \cdot [H^+_{sur}]}{[RCOOH]}. \tag{4.11}$$

The condition of the particles balance in the system, expressed in terms of concentration of surface-active anions ($RCOO^-$) and protons (H^+_{sur}), takes the form:

$$[RCOO^-] + \frac{[RCOO^-] \cdot [H^+_{sur}]}{K_{ion}} = n, \tag{4.12}$$

where n – the total number of molecules in monolayer.

It should be kept in mind that $n = 1/S_i$ and ionization of fatty acid molecules gives a negative charge to monolayer. Then the charge density inside the monolayer is given by the equation:

$$\sigma^- = -\frac{1}{S_i} e \frac{1}{1 + \dfrac{[H^+_{sur}]}{K_{ion}}}, \tag{4.13}$$

where

$$[H^i_{sur}] = 10^{-\left(pH + \frac{e\varphi_{sur}}{2,3kT}\right)}. \tag{4.14}$$

Introducing (4.13) into (4.5) and (4.9), one can get:

$$\varphi_b = \frac{e\delta}{\varepsilon_0\varepsilon_1 S_i} - \frac{2kT}{e} Arsh \frac{e}{S_i\left(1 + \dfrac{[H^+_{sur}]}{K_{ion}}\right) \cdot \sqrt{8INkT\varepsilon_0\varepsilon_2}} \tag{4.15}$$

and

$$\Pi_2 = \Pi_1 + \frac{2kT}{e}\sqrt{8INkT\varepsilon_0\varepsilon_2} \cdot \left[chsh^{-1} \frac{e}{S_i\left(1 + \dfrac{[H^+_{sur}]}{K_{ion}}\right) \cdot \sqrt{8INkT\varepsilon_0\varepsilon_2}} - 1 \right]. \tag{4.16}$$

It follows from (4.13) that when $[H^+_{sur}] \gg K_{ion}$ the value of σ tends to zero, and when $[H^+_{sur}] \ll K_{ion}$ the value of σ tends to $-(1/S_i)$, respectively.

In this case φ_b and Π correspond to the limiting state of monolayers, that is, neutral (dipole) and completely ionized state. By changing the pH, ionic strength and subphase composition at $S_i = const$ and $T = const$, one has an opportunity to observe the variation of the surface and dipole constituents of the boundary potential and surface pressure in different monolayers.

Experimental results and analysis

In our experimental work the chromatography homogeneous chemicals of stearic acid and its derivatives were studied. The monolayers were obtained in the Langmuir trough made from Teflon with the dimensions $350 \times 120 \times 15$ mm. The monolayers were formed according to standard procedure from chloroform solution of investigated substances (concentration 10 mg/ml). The subphase solutions were prepared from the chemicals of high-qualification grade using twice-distilled water. The boundary potential and the surface pressure in monolayers were measured by the method of dynamic capacitor and by the Wilhelmy method, respectively. The accuracy of measurements was about 2 mV and 0.1 mN/m, respectively. The experiments were carried out under isochoric conditions.

Figure 4.1a,b shows the isochoric plots of the boundary potential and the surface pressure versus pH and ionic strength of subphase for the monolayers of stearic acid with the area per molecule $S_i = 0.22$ nm². It is in accordance with close-packed state.

It is seen that isochors φ_b – pH have the well-defined asymmetric form with several sections. Within the pH interval from 1 to 3 (i.e., on the strongly acidic subphases), the surface potential does not depend upon pH and subphase composition. But at pH > 3 the magnitudes of φ_b become sensitive to pH. In particular, on the salt-free subphase (twice-distilled water + KOH) and on the subphase containing 0.1 M KCl, the isochors are divided into three sections with different slopes (BC, CD, DI). The points B, C, D (the curves 1, 2) correspond with the pH interval 3–4; 8.5 and 11. On the other hand if there is electrolyte of 2-1 type (CaCl₂) in the subphase, the sections BC and CD are merged into one monotonously falling branch (curve 3), and the point D shifts a little up and to the left in comparison with its position on a curve 2. In this case the section DI represents a rise of boundary potential.

Within the pH interval from 1 to 5 all isochors of surface pressure are merged and exhibit the horizontal shape. At pH > 5 the surface pressure becomes sensitive upon pH and ionic composition of subphase. Surface pressure and boundary potential exhibits a minimum in the pH range from 11 to 12.

Because the monolayers of fatty acids on the strongly acidic subphase exist in nondissociated (neutral) form, the shift of boundary potential to the negative side within the pH limits from 3 to 11 (curves 1,2) and from 3 to 10 (curve 3) can be explained by ionization of carboxyl groups of acidic

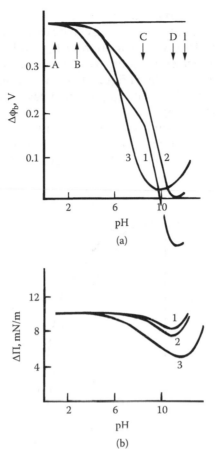

Figure 4.1 The experimental data of the boundary potential (a) and the surface pressure (b) versus pH at $S_i = 0.22$ nm²/molecule in the monolayers of stearic acid. 1–salt-free subphase; 2–0.1 M KCl; 3–40 mM CaCl₂.

molecules. Thus, the local minimum (points D) should correspond with the monolayers state, when the surface charges shielding by subphase electrolyte make a greater contribution than «accumulation» of additional surface charges at the expense of pH increasing.

However, two additional circumstances should be taken into account in the studied system: the presence of breaks at point C (Figure 4.1a, curves 1–2) and the existence of falling sections on the surface pressure isotherms (Figure 4.1b, curves 1–3). These are not in agreement with the abovementioned explanation because the theory predicts that the surface potential jumps within the pH limits from 1 to 11 should change smoothly and the values of Π_{sur} should increase monotonically as the value of σ grows (see equations 4.2 and 4.9).

To resolve this contradiction, the pH-induced variation of dipole constituents (i.e., the values of Π_{dip} and φ_{dip}) can be expected to occur in monolayers of fatty acid along with variation of Π_{sur} and φ_{sur} constituents. In particular, this is indicated by comparing experimental curves $\Delta\varphi_b$ – pH and Π – pH with calculated data (Figure 4.2). The curves were obtained for pH-dependent change of φ_{sur} and Π_{sur} for monolayers with $S_i = 0.22$ nm²/ molecule by equation (4.15).

It is seen that on salt-free and salt-containing subphases the actual changes of boundary potentials are in good agreement with the calculated data within the interval of pH 0–8.5. And within the pH interval

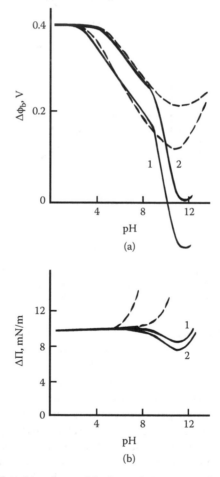

Figure 4.2 The calculated isochors of the boundary potential and the surface pressure (dotted lines) and experimental isochors (solid lines) versus pH at: 1–salt-free subphase, and 2–0.1 M KCl ($S_i = 0.22$ nm²/molecule).

from 9 to 12 the divergences become essential. Actually only abscissas of local minima coincide with theoretical and experimental curves (pH 11).

Figure 4.3 shows the change of «dipole» constituents $\Delta\Pi$ and $\Delta\varphi$ of stearic acid monolayers. They were obtained as the differences between the measured values of Π and φ_b and the calculated values of Π_{sur} and φ_{sur}. It is seen that the magnitudes of «dipole» components of Π and φ_b sharply drop at the certain values of pH that are dependent upon the structure of subphase electrolyte. In the case of water and 0.1 M KC1 the greatest deviations are observed within pH 8–11; and in the case of $CaCl_2$, within pH 5,5–8. The change of $\Delta\varphi$ and Π values is proportional to pH. This gives the basis to believe that the accumulation of molecules with «changed» dipoles occurs proportionally to logarithm of protons concentration in subphase.

The possible reasons of these effects can be the following factors: pH-dependent formation of ion–ionic and ion–molecular associates in monolayer, stabilized by hydrogen bonds; conformation change of fatty acid molecules at location sites of carboxylic groups; and also cross-linking in pairs of polar groups of amphiphilic molecules by cations of subphase [7,13,15,16]. It is difficult to identify the contribution of each of these factors. Therefore, we have compared the results obtained using monolayer of stearic acid with pH-dependent change of boundary potential jumps in monolayers of anilide of this acid and 1,5-bis(4-stearoilaminophenyl)-3-phenylformazane, of which the molecules have identical hydrocarbon radicals but have the polar heads with more rigid space structure. The chemical formulas of these compounds and schemes of their ionization are given below:

where $R = -(CH_2)_{16}-CH_3$.

Figure 4.4 shows the isochors φ_b – pH for condensed monolayers made from the mentioned substances. They have no anomalies peculiar to the monolayers of stearic acid. They are smooth and similar to calculated curves φ_{sur} – pH for the model systems, where the boundary potential is determined by pH-dependent surface charge and its screening by subphase electrolyte. Inasmuch as the change of the dipole component of the boundary potential was not detected for the monolayers of anilide and phenylformazane, it is possible to conclude that the changes of φ_{dip} and

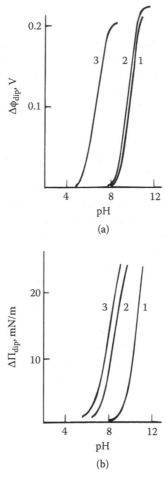

Figure 4.3 The variations of the dipole components of boundary potential (a) and the surface pressure (b) in monolayers of stearic acid versus pH at $S_i = 0.22$ nm^2/molecule. 1–salt-free subphase; 2–0.1 M KCl; 3–40 mM CaCl$_2$.

Π_{dip} in the monolayers of stearic acid are due to the pH-induced changing of the molecules' three-dimensional structure.

Interrelation between the boundary potential, surface pressure, and proton equilibrium in phosphatidylserine monolayers

Phosphatidylserine is a typical negatively charged amphiphatic lipid [2,3]. It is an obligatory component of cell membranes of animals, higher plants, and microorganisms [2]. In works on reconstitution of biophysical

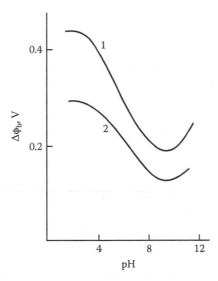

Figure 4.4 The isochors of the boundary potential of condensed monolayers of 1,5-bis(4-stearoilaminophenyl)-3-phenylformazane (1) and anilide of stearic acid (2) versus pH at salt-free subphase.

processes this lipid is used to form Langmuir monolayers with negatively charged surface faced to the subphase. And it is also used when it is necessary to modify phosphatidylcholine bilayers and liposomes in order to shift their surface charge toward more negative value. However, the lack of information on the structure of double electric layers in these systems is often conducive to incorrect interpretations of experimental results.

In this context, this section discusses the results of our investigation of boundary potential of phosphatidylserine monolayers as a function of the subphase pH, KCl concentration in it, and packing density of lipid molecules. Special attention is given to examining the relationship between the surface charge density of monolayers and shifts of proton equilibrium.

Following a standard technique described in [17], monolayers were formed from solutions (1 mg/ml) of commercial phosphatidylserine in chloroform in Teflon Langmuir cuvette (325 × 120 × 15 mm; volume 0.6 liters) at continuous subphase stirring at 20°C. Boundary potentials were measured by the dynamic capacitor method, accurate to ~2 mV. Surface pressure was measured according to the Wilhelmi method using a half-immersed rough platinum plate (0.1 × 45 × 10 mm) connected to electro-microbalance (sensitivity 0.1 mN/m). The pH of subphase was controlled by I-102 ionometer. Subphase solutions were prepared using KCl, KOH, HCl of special purity grade, and twice-distilled water.

The experimental procedure was as follows: The Langmuir cuvette was initially filled with twice-distilled water (pH 6). Then a monolayer of desired packing density was formed on the water surface. Thereafter, the subphase pH was adjusted to the required level by KOH or HC1 solutions. An analogous procedure was used in the experiments with monolayers on saline subphases; that is, the monolayers were initially formed on the surface of twice-distilled water, followed by KC1 (0.1 M) addition to the subphase and pH adjustment with KOH or HC1. In such a way the identity of conditions in the studied monolayer–subphase systems was achieved, allowing quantitative analysis of surface pressure and boundary potential jumps in response to external impacts.

The isochors of surface pressure and boundary potential versus pH for phosphatidylserine monolayers with two different packing densities of lipid molecules (S_i = 0.75 and 1.5 nm^2/molecule) obtained on saline and salt-free subphases are shown in Figure 4.5a,b.

In the first place of interest is the similarity of the general course of curves φ_b – pH and $\Delta\Pi$ – pH for different S_i for monolayers studied on saline subphase (Figure 4.5a) and their substantial differences for the monolayers studied on salt-free subphase (Figure 4.5b). Specifically, it is seen that on the saline subphases the boundary potential of monolayer decreases monotonously with the rise of pH from 1 to 12, whereas the surface pressure has a weakly expressed minimum within the pH limit from 1 to 2.

In the monolayers studied on salt-free subphases, a monotonous variation of $\Delta\varphi_b$ versus pH takes place only at high packing densities of molecules (S_i = 0.75 nm^2/molecule). Herewith, the surface pressure remains invariable in the pH range from 3 to 8, whereas outside this range it is increased. Regarding wider monolayers, the function $\Delta\varphi_b$ – pH in this case has a minimum at pH 8. Herewith, the surface pressure repeats in general the course of $\Delta\Pi$ – pH curves characteristic of monolayers in contact with the saline subphase. But the range of pH at which $\Delta\Pi \approx 0$ is within the limits from 2 to 6.

Based on the data on the structure of phosphatidylserine molecules containing phosphate, amine and carboxyl ionizable moieties, we have calculated the values of $\Delta\varphi_b$, Π and σ (surface charge density) depending upon pH and S_i for the model phosphatidylserine monolayers with packing densities of 0.5 and 1.5 nm^2/molecule, respectively.

We have proceeded from the following general equations describing surface pressure and boundary potential in the subphase–monolayer–air system [13]:

$$\varphi_b = \varphi_d + \varphi_s, \tag{4.17}$$

$$\Pi = \Pi_{kin} + \Pi_{disp} + \Pi_{dip} + \Pi_s, \tag{4.18}$$

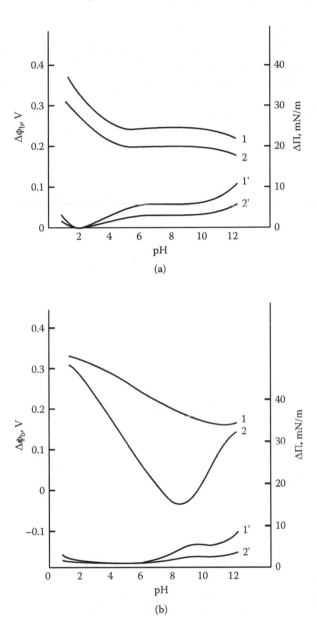

(a)

(b)

Figure 4.5 Effects of pH and specific area (S_i) on the steps of boundary potential (1, 2) and surface pressure (1′, 2′) in phosphatidylserine monolayers. a) on the
saline subphase (0.1 M KCl); b) on the salt-free subphase. 1 and 1′ - S_i = 0.75 nm²/
molecule; 2 and 2′ - S_i = 1.5 nm²/molecule.

where φ_d – dipole component of boundary potential, φ_s – surface component of boundary potential, Π_{kin} – kinetic component of two-dimensional pressure, Π_{disp} – Van der Waalsian component of two-dimensional pressure, Π_{dip} – dipole component of two-dimensional pressure, and Π_s – surface charge component of two-dimensional pressure.

Regarding the object under consideration, the first and the second terms of equation (4.18) may be disregarded because at constant specific area in monolayers (isochoric conditions) these constituents make no contributions to the variation of Π and φ_b. And in the first approximation the values of φ_{dip} and Π_{dip} in (4.17) and (4.18) may be assumed to be constant. In such a case the last members in equations (4.17) and (4.18) can be represented as follows [13]:

$$\varphi_s = \frac{2kT}{e} Arsh \frac{\sigma}{\sqrt{8INkT\varepsilon_0\varepsilon_2}}, \tag{4.19}$$

$$\Pi_s = \frac{2kT}{e} \sqrt{8INkT\varepsilon_0\varepsilon_2} \cdot \left(chsh^{-1} \frac{\sigma}{\sqrt{8INkT\varepsilon_0\varepsilon_2}} - 1 \right), \tag{4.20}$$

where e – elementary charge, k – Boltzmann constant, T – absolute temperature, σ – surface charge density, I – subphase ionic strength, N – Avogadro's number, ε_0 – absolute dielectric constant, and ε_2 – subphase dielectric constant.

The surface charge density in equations (4.19) and (4.20) can be described based on the concepts of ionization of the phosphate, carboxyl, and amine groups in the phosphatidylserine monolayer, that is:

$$RPO^- + H^+ \leftrightarrow RPOH \tag{4.21}$$

$$RCOO^- + H^+ \leftrightarrow RCOOH \tag{4.22}$$

$$RNH_2 + H^+ \leftrightarrow RNH_3^+ \tag{4.23}$$

The possible binding of metal cations with phosphate and carboxyl groups can be accounted by the following equations:

$$RPO^- + Me^+ \leftrightarrow RPOMe \tag{4.24}$$

$$RCOO^- + Me^+ \leftrightarrow RCOOMe \tag{4.25}$$

So, the expressions for the equilibrium ionization constants and the reactions of salt formation take the following form:

$$K_p = \frac{\left[RPO^-\right] \cdot \left[H_s^+\right]}{\left[RPOH\right]}. \qquad (4.26)$$

$$K_A = \frac{\left[RCOO^-\right] \cdot \left[H_s^+\right]}{\left[RCOOH\right]}. \qquad (4.27)$$

$$K_B = \frac{\left[RNH_2\right] \cdot \left[H_s^+\right]}{\left[RNH_3^+\right]}. \qquad (4.28)$$

$$K_{p.ads} = \frac{\left[RPO^-\right] \cdot \left[Me_s^+\right]}{\left[RPOMe\right]}. \qquad (4.29)$$

$$K_{s.ads} = \frac{\left[RCOO^-\right] \cdot \left[Me_s^+\right]}{\left[RCOOMe\right]}. \qquad (4.30)$$

Having the balance of surface and bulk concentrations of all components of the monolayer–subphase system similarly to [18], one can obtain an expression for monolayer surface charge density versus concentrations of inorganic ions and equilibrium constants for reactions (4.21) – (4.25):

$$\sigma = \frac{1}{S_i} e \left[\frac{1}{1 + \dfrac{K_A}{\left[H_s^+\right]}} - \frac{1}{1 + \dfrac{\left[H_s^+\right]}{K_P} + \dfrac{\left[Me_s^+\right]}{K_{p.ads}}} - \frac{1}{1 + \dfrac{\left[H_s^+\right]}{K_A} + \dfrac{\left[Me_s^+\right]}{K_{s.ads}}} \right]. \qquad (4.31)$$

where $[H_s^+] = 10^{-pH} \exp(-e\varphi_s/kT)$ – surface concentration of protons in the monolayer–subphase system, and $[Me_s^+] = [Me_v^+]\exp(-e\varphi_s/kT)$ – surface concentration of metal ions.

Figure 4.6 shows the results of calculation of surface pressure, boundary potential, and surface charge density as the functions of pH and S_i at C = 0.1 M obtained using equations (4.19), (4.20), and (4.31). The values of K_A, K_B, $K_{p.ads}$, $K_{a.ads}$, and φ_d were selected so that calculated functions coincided to the maximum with experimental data. Comparing these data with the graphs presented in Figure 4.5 shows that the calculated functions of φ_b and Π_s upon pH virtually repeat the course of experimental curves φ_b – pH and $\Delta\Pi$ – pH at both monolayer packing densities

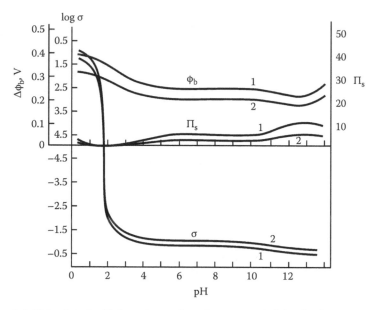

Figure 4.6 Calculated pH functions of surface potential (φ_{sur}), electrostatic part of surface pressure (Π_s) and surface charge density (σ) for the phosphatidyl-serine monolayers with different specific areas in saline subphase (C = 0.1 M). $1-S_i$ = 0.75 nm^2/molecule; $2-S_i$ = 1.5 nm^2/molecule.

(S_i = 0.75 and 1.5 nm^2/molecule). Thus, the model with parameters $K_A = 1 \times 10^{-2}$, $K_B = 1 \times 10^{-9}$, $K_P = 1 \times 10^{-1}$, $K_{p.ads} = 1 \times 10^1$, $K_{a.ads} = 1 \times 10^{30}$ M (no adsorption), φ_d = 0.35 V at S_i = 0.75 nm^2/molecule, and φ_d = 0.29 V at S_i = 1.5 nm^2/molecule reflects adequately the behavior of a real mono-layer on a saline subphase. In this case the course of function of σ upon pH indicates that at pH < 1.5 the monolayers have a positive surface charge, and a negative one at pH > 1.5. The surface charge density varies dramatically in the pH range from 1 to 4. At pH from 5 to 11 it is about 3.9×10^{-2} C/m^2 and reaches 6.3×10^{-2} C/m^2 at pH 12. The dipole potential in the monolayers contacting saline subphase is independent of pH and amounts to 0.35 V at S_i = 0.75 nm^2/molecule and 0.29 V at S_i = 1.5 nm^2/ molecule. This corresponds to effective dipole moments 9.3×10^{-30} C × m and 1.5×10^{-29} C × m, if the relative dielectric constant in monolayers is equal to 4.

Hence, the phosphatidylserine monolayers on saline subphases may have both positive and negative surface charges. In the presence of 0.1 M KCl the point of zero charge in these monolayers is at pH 1.5.

Figure 4.7 shows the calculated isochors of Π_s, $\Delta\varphi_b$, and σ as the func-tions of pH for monolayers studied on salt-free subphase. Comparing these data with the data shown in Figure 4.5, one can see that the model with parameters $K_A = 1 \times 10^{-2}$, $K_B = 1 \times 10^{-9}$, $K_P = 1 \times 10^{-1}$, $K_{p.ads} = 1 \times 10^1$ M,

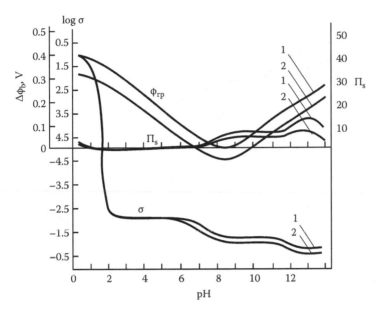

Figure 4.7 Calculated isochors of boundary potential steps (φ_b), electrostatic component of two-dimensional pressure (Π_s) and surface charge density (σ) versus pH for phosphatidylserine monolayers with different specific areas in salt-free subphase. $1–S_i = 0.75$ nm^2/molecule; $2–S_i = 1.5$ nm^2/molecule.

$\varphi_b = 0.29$ V and 0.35 V fits the experimental isochor only at $S_i = 1.5$ nm^2/molecule (expanded monolayer). However, at lower values of specific area the results of calculations at indicated parameters do not show a correlation with the experimental data. The coincidence of the general course of theoretical and experimental functions $\Delta\varphi_b$ – pH and $\Delta\Pi$ – pH at $S_i = 0.75$ nm^2/molecule is achieved only when $K_{p.ads}$ and φ_b are equal to 1×10^{-2} M and 0.25 V, respectively (see Figure 4.8). This result suggests that at increased densities of molecule packing and at low ionic strength of subphase, the conformation of the phosphatidylcholine polar groups is changed and their phosphate groups become inaccessible to the ions of subphase. If this is indeed the case, the lipid molecules in fact lose one of their negative charges. In this case the general course of calculated function σ – pH shows that at pH < 5.5 the phosphatidylserine monolayers are positively charged and become negatively charged at pH > 5.5. Respectively, the density of pH-induced charge in the physiological pH range becomes 100–1000 times smaller than what follows from the concept that each lipid molecule carries two free negative charges and one positive charge [2,3]. In our opinion, this circumstance is important and must be accounted in studies using phosphatidylserine as negatively charged lipid.

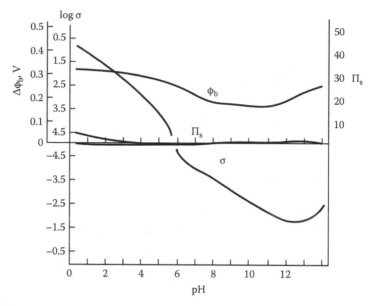

Figure 4.8 Calculated pH dependences of boundary potential steps (φ_b), electrostatic components of two-dimensional pressure (Π_s), and surface charge density (σ) for the phosphatidylserine monolayers with $S_i = 0.75$ nm^2/molecule obtained considering changes of φ_b and screening of one acidic group in the polar heads of the lipid molecules. For the model parameters see the text.

Investigation of the dipole component changes of the boundary potential jumps in phosphatidylcholine monolayers

We have also studied the influence of chemical modification of phosphatidylcholine molecules on the boundary potential jumps in monolayers formed at the air–water interface. It was shown that the reduction of a carbonyl group in dipalmitoylphosphatidylcholine molecules to CH_2 leads to a decrease in the boundary potential jumps approximately by 100–130 mV. The reduction of both carbonyl groups results in a decrease of these values by approximately 180–200 mV. The changes in the potential jumps make 20 and 33% of those in nonmodified monolayers.

Phosphatidylcholine monolayers are widely used in studies and reconstitution of various biophysical processes in biological membranes [8,19–24]. Ordered oriented hydrocarbon radicals of phosphatidylcholine molecules simulate a hydrophobic zone of biological membrane while the polar heads simulate their external hydrophilic surface. The physical–chemical properties of these monolayers depend on three-dimensional structure of lipid molecules, their packing density at the interface, pH of the subphase, and its ionic composition and concentration. These factors affect the magnitudes of two-dimensional pressure and boundary potential in the monolayer–subphase systems. In analyzing these parameters under various experimental conditions, one can understand rather fine functioning peculiarities of lipid matrix in native membrane structures and their synthetic analogues [25–31].

We have considered the basic regularities of the boundary potential in the monolayers of dipalmitoylphosphatidylcholine (DPPC) in [18,31]. These works were devoted to study of the effect of pH and ionic subphase composition on the ionization degree of phosphate and choline fragments in the polar heads of lipid molecules.

Below, the effect of chemical modification of phosphatidylcholine molecules on the dipole part of the boundary potential jumps is described. In this context, we have studied the variations of the boundary potential upon pH and ionic composition of the subphase in the monolayers of DPPC, l-palmitoyl-2-hexadecylphosphatidylcholine (PHPC), l-hexadecyl-2-palmitoylphosphatidylcholine (HPPC), and di-hexadecylphosphatidylcholine (DHPC).

Monolayers were formed from solutions of commercial samples of DPPC (Sigma, USA) and their synthetic chromatographically homogeneous analogues (PHPC, HPPC, DHPC) at concentrations 0,1 mg/ml in chloroform in teflon Langmuir trough (325 × 120 × 15 mm, 0.6 liters) at 20°C and subphase stirring.

The boundary potential was measured by the dynamic capacitor method accurate to 2 mV. The two-dimensional pressure was registered by the Wilhelmi method using a semi-immersed platinized platinum plate (0.1 × 45 × 10 mm) connected with electro-microbalance Sartorius (sensitivity 0.1 mN/m). Subphase pH was checked by ionometer I-102. Subphase solutions were prepared from reagents (KC1, KOH, HCl, CaCl$_2$) of special purity grade and twice-distilled water. The analyzed data were the compression isotherms of monolayer and replotted isochoric functions of the boundary potential jumps upon pH, ionic composition of the subphase, and packing density of lipid molecules.

It was found that the initial packing density of lipid molecules (S_i) in DPPC monolayers formed on salt-free subphases with certain values of pH significantly affects the dynamics of boundary potential jumps in these monolayers at their subsequent compression. Figure 4.9 shows the compression isotherms of DPPC monolayers obtained on subphases with pH 6, 9, 10, and 12, when the initial values of S_i were, respectively, 0.6 nm^2/molecule (condensed monolayer) and 1.2 nm^2/molecule (initially expanded monolayer). As seen, the curves $\Pi - S_i$ upon pH are coincided in the both cases, while the curves $\varphi_b - S_i$ are differed. At any values of the specific areas (S_i) within the compression range, the values of φ_b upon pH in the initially condensed monolayers are decreased monotonically, while in the initially expanded monolayers they fall nonmonotonically, that is, had a local extreme.

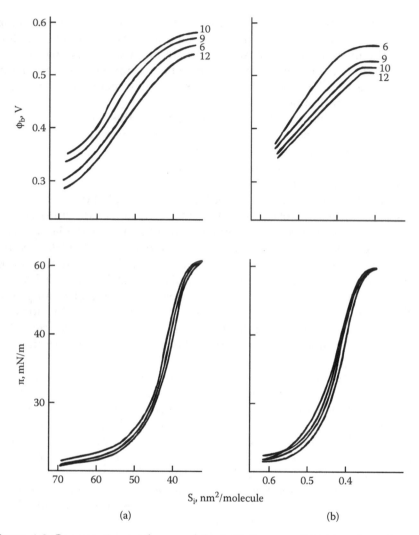

Figure 4.9 Compression isotherms of the initially expanded (a) and condensed (b) monolayers of DPPC deposited on the salt-free subphases with different pH (pH values are shown at the curves).

The isochors of the boundary potential as a function of pH for the extreme packing densities in the initially expanded and initially condensed monolayers of DPPC, obtained by crossing the compression isotherms of these monolayers at $S_i = 0.4$ nm²/molecule, are shown in Figure 4.10. The curved shape gives a general picture of the monolayer response to the shift of pH depending on S_i. In particular, in the initially condensed monolayer of DPPC, the boundary potential is decreased gradually as pH rises; and in the initially expanded one, φ_b pass through a maximum at pH 10.

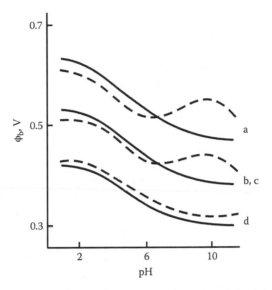

Figure 4.10 Isochors of the boundary potential jumps of the initially expanded monolayers (broken lines) and condensed monolayers (solid lines) of DPPC (a), PHPC and HPPC (b,c), and DHPC (d) at $S_i = 0.4$ nm^2/molecule investigated in salt-free subphases with different pH.

The isochors φ_b – pH for condensed monolayers of DPPC were analyzed in [18]. It was shown that monotonic decrease of the boundary potential caused by the rise of subphase pH reflects the accumulation of negative charges in the plane of the polar heads of lipid molecules. The dipole component of the boundary potential in that monolayer was observed constant. The nonmonotone shape of the boundary potential, which was observed on the initially expanded monolayer of DPPC, is unexpected. This effect is apparently due to the pH-induced change of the dipole moments in lipid molecules. It manifests itself when the initial interaction of lipid molecules with subphase is realized at the excess of free area.

Taking into account that the main dipole-determining fragments of phosphatidylcholine molecules are C = O sites, we have carried out a series of experiments on the compression of initially condensed and initially expanded monolayers consisting of PHPC, HPPC, and DHPC. They differ from DPPC by one or both carbonyl groups reduced to CH$_2$. As seen in Figure 4.10 (curves b, c), the isochors φ_b – pH for monolayers of PHPC and HPPC on salt-free subphases have the same shape as for DPPC (curve a). As compared with the case of DPPC monolayers, the values of φ_b in all investigated ranges of pH decreased by 0.1–0.12 V. However, in DHPC monolayers (Figure 4.10d), one failed to find any effect of the initial packing density of amphiphilic molecules on the shape of $\varphi_b - pH$ curves. Thus, in contrast with DPPC, PHPC, and HPPC monolayers, DHPC monolayer is insensitive to the initial density of molecule packing. The values of φ_b in this monolayer at pH 6 amount to 0.37 V at $S_i = 0.4$ nm²/molecule. This is 0.08–0.1 V less than in monolayers of PHPC and HPPC, and 0.2–0.22 V less than in DPPC monolayer.

The above results suggest that the structural rearrangement occurs on salt-free subphases in phosphatidylcholine monolayers whose molecules have at least one carbonyl group. This rearrangement depends on the initial condition of monolayer formation. In the presence of additional electrolyte in the subphase (sodium or calcium chlorides with ionic strengths about 0.1 g–eq/l) the boundary potential in DPPC, HPPC, and PHPC monolayers become insensitive to the initial packing density of their molecules. The shapes of the $\varphi_b - pH$ isochors within all ranges of subphase pH prove to be smooth and similar to the $\varphi_b - pH$ isochors for DHPC monolayers in whose molecules both carbonyl groups are reduced to CH_2. At the same time, the differences in the values of φ_b determined by specific structure of DPPC, HPPC, and DHPC monolayers remain.

The isochores $\varphi_b - pH$ for DPPC, PHPC, HPPC, and DHPC monolayers compressed on salt-containing subphases are shown in Figure 4.11. The comparison of Figure 4.10 and Figure 4.11 shows that sodium and calcium chlorides totally eliminate the course anomalies of the boundary potential jumps. The reason is as follows: K⁺ and Ca^{2+} have affinity to ionized fragments of lipid molecules.

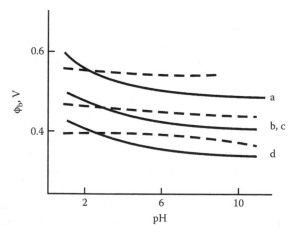

Figure 4.11 Isochors of the boundary potential jumps of the monolayers of DPPC (a), PHPC and HPPC (b,c), and DHPC (d) at $S_i = 0.4$ nm²/molecule obtained on subphase with ionic strength of 0.1 g–eq/l in the presence of KCl (broken lines) or $CaCl_2$ (solid lines).

On the neutral salt-free subphases (pH 6,5) the charges of phosphate and choline fragments in phosphatidylcholine molecules almost totally compensate each other [15]. Acidic and alkaline subphases make pH-proportional contribution into the boundary potential jumps. It is natural to assume that the differences in the values of φ_b associated with the initial conditions of the existence of DPPC, PHPC, and HPPC monolayers on alkaline subphases are due to pH-induced change of dipole moments in their molecules. It is noteworthy that pH-dependent transformation of dipole moments in phosphatidylcholine molecules requires at least one nonreduced carbonyl group, low ionic strength of subphase and initially expanded state of monolayer, when the initial specific area in the boundary film is significantly higher than the cross-section area of the lipid molecule.

When the excess of electrolyte is present in the subphase (KCl or $CaCl_2$ with ionic strengths about 0.1 g–eq/l) their ions appear to screen free and dipole charges of phosphatidylcholine molecules so strongly that they prevent conformational change. Analysis of the experimental data according to the Helmholtz model [13] enables one to make some estimation. Thus, using the equation

$$\varphi_d = \frac{e_0 \delta}{\varepsilon S_i} \qquad (4.32)$$

where e_0 – elementary charge, δ – dipole distance in the monolayer space, ε – dielectric constant in the monolayer, and S_i – specific area.

It is easy to find the value of δ/ε for the point of zero charge of the lipid monolayers depending on the structure of their molecules and to determine the change of δ/ε under the influence of the experimental conditions.

For monolayers DPPC, PHPC, HPPC, and DHPC investigated on neutral salt-free subphase (pH 6.5), the values of δ/ε at $S_i = 0.4$ nm^2/molecule are 1.37, 1.05, 1.0, 0.825 V×m^2/K, respectively. Therefore, the replacement of one of the carbonyl groups in phosphatidylcholine molecules by CH_2 group leads to 1.3-fold decrease in the effective dipole moment of monolayer. The replacement of both carbonyl groups leads to 1.66-fold decrease. At pH 10 the differences in the values of δ/ε for the initially expanded and initially condensed monolayers of DPPC, PHPC, and HPPC reach 0.225 and 0.2 V × m^2/K. In other words, a shift in subphase pH by 3.5 to the alkaline side (relative to pH 6.5) determines a 1.1-fold increase in the effective dipole moments for the initially expanded DPPC monolayer. For the initially expanded monolayers of PHPC and HPPC, it is 1.12-fold.

As for monolayers DPPC, PHPC, HPPC, and DHPC investigated in the presence of KCl and $CaCl_2$, the respective isochors show that the dipole moments in their molecules do not depend on the initial densities of monolayer packing. They are determined only by the peculiarities of the structure of the lipid molecules. The reduction of one or both carbonyl groups to the CH_2 group leads to the same decrease of δ/ε as for the monolayers on neutral salt-free subphase.

All the above data suggest that the studied monolayers are significantly more complex systems than was previously believed. The boundary potential jumps in them at a constant packing density of amphiphilic molecules depend both on the ratios of the ionization of the phosphate and choline fragments in the polar heads and on the peculiarities of the chemical structure of the fragments binding these heads to the hydrocarbon radicals. In particular, the presence of one or both carbonyl oxygens in the phosphatidylcholine molecules makes it possible for them to change conformation under certain conditions (small ionic strength of the subphase, significant specific area, and the required level of pH). This should be taken into account in operation with both phosphatidylcholine monolayers and bilayers used as models of biological membranes.

References

1. MTP International Review of Science. Biochemistry. Ser. 1. V. 2. *Biochemistry of Cell Walls and Membranes*. Ed. Fox C. F. London: Butterworth; Baltimore: University Park Press, 1975.
2. Gurr M.I., James A.T. *Lipid Biochemistry: An Introduction*. London, New York: Chapman and Hall, 1980.
3. Lehninger A. Biochemistry. Molekulyarnie osnovi strukturi i phunktsii kletki / translated from Engl. M.: Mir, 1974. 956 p. (In Russian).
4. Ksenzhek O.S., Gevod V.S. *Surfaktanti legkogo v norme i patologii*. Kyiv: Nauk. dumka, 1983. P. 155–163.
5. Ksenzhek O.S., Gevod V.S., Aianyan A.E., Miroshnikov A.I. Bioorgan. *Khimia*. 1981. 7(11):1680–1687. (In Russian).
6. Small D.M. *Handbook of Lipid Research*. 1986. 4:285–343.
7. Haines T.H. *Proc. Nat. Acad. Sci. USA*. 1983. 80:160–164.
8. Chapman D. *Biological Membranes, Physical Fact and Function*. London, New York: Acad. Press, 1968.
9. Hargreaves W.R., Deaner D.W. *Biochemistry*. 1978. V. 17. № 18. P. 3759–3767.
10. Matijevic E., Pethica B.A. *Trans. Faradau Soc*. 1958. V. 54. P. 1382–1407.
11. Monolayers. *Advances in Chemistry Series*. № 144. Washington: Amer. Chem. Soc., 1975. 230 p.
12. Tomoaia-Cotisel M., Zsako J., Mocanu A., Lupea M., Chifu E. *J. Colloid and Interface Sci*. 1987. V. 117. № 2. P. 464–476.
13. Gaines G.L. (Jr.) *Insoluble Monolayers at Liquid–Gas Interfaces*. N.Y.: Wiley-Interscience, 1966. 386 p.
14. Gevod V.S., Ksenzhek O.S., Solov'ev E.L. *Biol. Membrany*. 1990. V. 7. N 4. P. 428–434. (in Russian).
15. Ksenzhek O.S., Gevod V.S. Dvumernoie davlenie i strukturnaya organizatsiya monosloev. M., 1975. - Dep. in VINITI N 1657–75. (In Russian).
16. Boggs J.M. *Biochim. et biophys. acta*. 1987. V. 906. P. 353–404.

17. Adamson, Fizicheskaya Khimiya Poverkhnostey (*Physical Chemistry of Surfaces*) (Moscow: Mir, 1979): 568 p. (Russian translation).
18. Gevod, V. S., O. S. Ksenzhek, I. L. Reshetnyak, and E. L. Solov'ev. / *Biol. Membrany.* 1990. V.7, p.779–787. (In Russian).
19. Tien H. T. K. and Jr. James, *Chemistry of Cell Interface.* Part A. Chapter IV (New York: Acad. Press, 1971): 205.
20. Mobius, D. Ber. Bunsenges. *Phys. Chem.* 82:848–858 (1978).
21. Helm, C. A., H. Mohwald, K. Kjaer, and J. Als-Nielsen, *Europhysics Lett.* 4:697–703 (1987).
22. Miller, C. A., Helm, and H. Mohwald, *J. Physique* 48:693–701 (1987).
23. Miller, C. A., W. Knoll, and H. Mohwald, *Phys. Rev. Lett.* 56:2633–2636 (1986).
24. Helm, C. A., H. Mohwald, K. Kjaer, and J. Als-Nielsen, *Biophys. J.* 52:381–390 (1987).
25. Gevod, V. S., O. S. Ksenzhek, and A. I.-Miroshnikov, Biomembrany. Struktura, Funktsii, Meditsinskie Aspekty (*Biomembranes. Structure, Function, Medical Aspects*) (Riga: Zinatne, 1981): 108–132. (In Russian).
26. Gevod, V. S., O. S. Ksenzhek, and I. L. Reshetnyak / *Biol. Membrany.* 1991. V.8. p.423–429. (In Russian).
27. Gevod, V. S., O. S. Ksenzhek, and E. L. Solov'ev. / *Biol. Membrany.* 1990. V.7. p.428–434. (In Russian).
28. Birdi, K. S. *Proceedings of 7th Scandinavian Symposium on Surface Chemistry* (Denmark, 1981): 103–127.
29. Pethica, J. Mingins, and J. A. G. Taylor, *J. Colloid Interface Sci.* 55:2–8 (1976).
30. Hayashi, M., T. Muramatsu, and I. Hara, *Biochim. Biophys. Ada* 255:98–106 (1972).
31. Ksenzhek O. S. and V. S. Gevod, *Proceedings of 6th USSR-Japanese Seminar on Electrochemistry.* The Fundamental Problem of Interface Structure and Electrochemical Kinetics (Hokkaido University Press, 1988): 77.

chapter five

Applications of interfacial electrical phenomena
Colloidal systems (charge behavior)

K. S. Birdi

Contents

Introduction

It will be useful to discuss certain specific applications where the *interfacial electrical phenomena* are known to be of importance. The subject of general chemistry also includes *electrochemistry* in the bulk phase. However, experiments have shown that the electrical charges are asymmetric near interfaces. One also finds in everyday life many important systems where these interfaces are present. One such system is where small particles—*colloids*—are present.

The electrical interfacial phenomena of various systems have been extensively described in the current literature. It is thus important to consider a few important examples of such systems. Since one finds that there are a very large number of such application areas, in this book only some of the most important systems will be described.

It is useful to describe the very large industrial application of colloid chemistry. Mankind has been aware of colloids for thousands of years.

Ancient civilizations such as those of Egypt and Maya used their knowledge about adhesion (between blocks of stones) when building pyramids and other tall structures (pillars, etc.). This was long (more than 2000 years) before modern-day cement (which basically consists of very fine particles—colloids) was invented.

Colloids

In everyday life one comes across solid particles of different sizes, ranging from sand particles or talcum powder to aerosols and dust floating around in the air. A special relation exists between size of particles (surface area) and their characteristics. The rather small particles (Figure 5.1) in the range of size from 50 Å to 50 μm are called **colloids**.

This can be explained if one considers what happens when sand particles are thrown into air versus dust particles. If one throws these two different particles in air, one notices that dust or other fine particles remain in suspension in air for a very long time, whereas the larger sand particles fall toward the earth very fast. The reason for this difference arises from the following analyses by Brown. In fact, once in a while one observes that a particle gets a collision-like thrust. In the 19th century, Brown observed under microscope that small microscopic particles suspended in water made some erratic movements (as if hit by some neighboring molecules). This has since been called *Brownian motion*. The erratic motion arises from the kinetic movement of the surrounding water molecules. Thus, colloidal particles would remain suspended in solution through Brownian motion only if the gravity forces did not drag these to the bottom (or top).

It is well known that sand particles very quickly fall to the earth if one throws these into the air. On the other hand, one finds that talcum particles stay floating in the air for a long time. This phenomenon is characteristic for small-sized particles. Particles are characterized according to size as shown in Table 5.1.

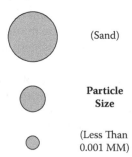

(Sand)

**Particle
Size**

(Less Than
0.001 MM)

Figure 5.1 Large (sand) and small (micrometer size) colloid particles (schematic) (μm = MM).

Table 5.1 Particle Characterization according to Size

Colloidal Dispersions	Radius (1000 μ to nm)
Mist/fog	0.1 μ–10 μ
Pollen/bacteria	0.1 μ–10 μ
Oil in smoke/exhaust	1 μ–100 μ
Virus	10 μ–100 μ
Polymers/macromolecules	100 nm–0.1 nm
Micelles	10 nm–0.1 nm
Vesicles	1 μ–1000 μ

The size of particles one encounters in various everyday systems is found to vary from 1000μ (1000 10^{-6} m) to nm (10^{-9} m) (nanoparticles) scale. The latter has introduced a modern nanotechnology and requires an additional dimension in such small particle science. Modern methods of observing these particles are carried out by the following apparatuses:

ordinary microscope (μ range)
electron microscope (nm range)
scanning probe microscope (SPM) (nm range)

The main characteristic of any system consisting of such colloidal particles concerns the *stability*—something that may be compared to whether the system will remain energetically stable or will take up a new state of more stable configuration. One may very roughly compare this to an object that is *stable* when standing up, but if tilted beyond a certain angle, it topples and comes to rest on its side (Figure 5.2).

A colloidal suspension may be *unstable* and exhibit separation of particles within a very short time. Or it may be *stable* for a very long time, such as over a year. And there will thus be found a *metastable* state, which would be between these two (such as in the case of mayonnaise). Further, there are also examples where a stable system becomes unstable under specific conditions (such as wastewater treatment). This is an oversimplified example, but it shows that one should proceed to analyze any colloidal system following these three criteria.

As an example, one may consider the wastewater treatment process. The wastewater with colloidal particles is a stable suspension. However, by treating it with some definite methods (such as pH control, electrolyte concentration, etc.), one can change the stability system as shown in Figure 5.3. The suspended particles aggregate and drop out of the wastewater. The clear water is filtered and treated with other processes, such as chlorine, before being pumped into the ocean or river.

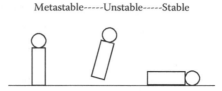

Figure 5.2 Stability criteria of any colloidal system: metastable—unstable—stable states.

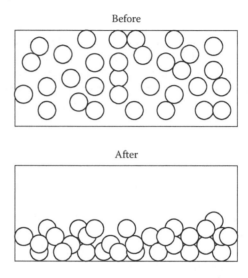

Figure 5.3 The suspension of particles is treated in the wastewater plant: incoming wastewater (before); unstable system after treatment (after) with suitable coagulants and so forth.

Another common example is that of an **emulsion** (mixture of oil and water and suitable emulgators). The emulsion can be—

very unstable or
very stable,

depending on the composition.

These different systems under consideration exhibit characteristics related to two main kinds of interactions (attractive forces—repulsive forces).

van der Waals forces: The universe is under different kinds of forces. All condensed matter is stabilized under specific forces. In colloidal systems the Waals forces play an important role. When any two particles (neutral or with charges) come very close to each other, the van der Waals forces will be strongly dependent on the medium surrounding; in a vacuum two identical particles will always exhibit attractive force.

Electrostatic forces

On the other hand, if two different particles are present in a medium (in water), then there may be present repulsion forces. This can be due to one particle adsorbing with the medium more strongly than with the other particle. One example is silica particles in water medium and plastics (as in wastewater treatment). It is important to understand under what conditions it is possible for colloidal particles to remain suspended. Especially, if paint aggregates in the container, then it is obviously useless.

When solid (inorganic) particles are dispersed in aqueous medium, ions are released in the medium. The ions released from the surface of the solid are of opposite charge. This can be easily shown when glass powder is mixed in water, and one finds that conductivity increases with time. The presence of the same charge on particles in close proximity gives repulsion, which keeps the particles apart (Figure 5.4).

The *positive–positive* particles will show repulsion. On the other hand, the *positive–negative* particles will attract each other. The ions' distribution will also depend on the concentration of any counter-ions or co-ions in the solution. Experiments show that even glass, when dipped in water, exchanges ions with its surroundings. Such phenomena can be easily investigated by measuring the change in conductivity of the water.

The electrical state of any surface (solid or liquid or macromolecule (such as protein; DNA) or bacteria or virus) depends on the spatial distribution of free (electronic or ionic) charges in its vicinity (varying from molecular diameters to um). This asymmetry of charge distribution is idealized as an electrochemical double layer (**EDL**). The same is found around micellar structures or polyelectrolyte molecules (DNA). The

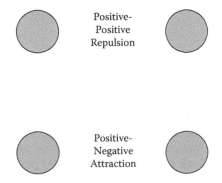

Figure 5.4 Solid–solid interactions with charges (positive–positive or positive–negative).

current theoretical analyses of **EDL** are based on a physical model in which the following items are classified:

1. One layer is assigned as a fixed charge or surface charge attached to the particle or solid surface.
2. The other layer is distributed more or less diffusely in the surrounding liquid media in contact with the given particle.

In the **second** layer one finds an excess of counter-ions, opposite in sign to the fixed charge. Actually, in this region is a deficit of co-ions of the same sign as the fixed charge. Counter- and co-ions in immediate contact with the surface are found to be located in the Stern layer. This situation is similar to a molecular capacitor. All the ions that are farther away from the surface form the diffuse layer, or Gouy layer. This is expected when one has to consider the surrounding water molecules.

The force, F_{12}, acting between these opposite charges is given by Coulombs law, with charges q_1 and q_2, separated at a distance R_{12}, in a dielectric medium, D_e:

$$F_{12} = (q_1 \, q_2)/(4 \, \pi \, D_e \, \varepsilon_o \, R_{12}) \tag{5.1}$$

The force would be attractive between opposite charges but repulsive in the case of similar charges. Since D_e of water is very high (80 units) as compared with D_e of air (ca. 2), we will expect very high dissociation in water, yet hardly any dissociation in air or organic liquids. Let us consider the F_{12} for Na^+ and Cl^- ions (with charge of $1.6 \; 10^{-19}$ C $= 4.8 \; 10^{-10}$ esu) in water ($D_e = 74.2$ at $37°$ C), and at a separation (R_{12}) of 1 nm:

$$F_{12} = (1.6 \; 10^{-19}) \, (1.6 \; 10^{-19})/$$

$$[(4 \; \Pi \; 8.854 \; 10^{-12}) \, (10^{-9})(74.2)] \tag{5.2}$$

$$= - \, 3.1 \; 10^{-21} \; \text{J/molecule}$$

where ε_o is $8.854 \; 10^{-12} \, kg^{-1}m^{-3} \, s^4 \, A^2$ ($J^{-1} \, C^2 \, m^{-1}$). This gives a value of F_{12} of $-3.1 \; 10^{-21}$ J/molecule or -1.87 kJ/mole.

Another very important physical parameter one must consider is the *size distribution* of the colloids. A system consisting of particles of same size is called a mono-disperse. A system with different sizes is called poly-disperse. Additionally, one must consider the role of shape (and shape variation distribution) in this context. It is also obvious that systems with mono-dispersion will exhibit different properties than those of poly-dispersion. In many industrial applications (such as coating on tapes used for recording music, coatings on CDs or DVDs) the size distribution becomes very important.

The methods used to prepare mono-disperse colloids is to achieve a large number of critical nuclei in a short interval of time. This induces all equally sized nuclei to grow simultaneously and thus produce a mono-disperse colloidal product.

Characteristics of colloids (DLVO theory)

The question one needs to understand is under which conditions a colloidal system will remain *dispersed* and under which conditions it would become *unstable*.

How colloidal particles interact with each other is one of the important questions that determines the understanding of the experimental results for phase transitions in such system as found in various industrial processes. One also will need to know under which conditions a given dispersion will become unstable (coagulation). For example, one needs to apply coagulation in wastewater treatment such that most of the solid particles in suspension can be removed.

Any two particles, which when they come close to each other, there will exist different forces:

Attractive Forces—Repulsive Forces

If the *attractive forces* are larger than the *repulsive forces*, then the two particles will merge together. However, if the repulsion forces are larger than the attractive forces, then the particles will remain separated (Figure 2.8).

It is important to mention here that the medium in which these particles are present will to some degree contribute also, especially such as pH and ionic strength (i.e., concentration of ions) are found to exhibit very specific effects.

The different forces of interest are:

van der Waals
electrostatic
steric
hydration
polymer–polymer interactions (if polymers are involved in the system)

In many systems one may add large molecules (macromolecules: polymers), which when adsorbed on the solid particles will impart a special kind of stability criteria. Merely the large size imparts steric hindrances for particles to approach closer, besides other characteristics.

It is well known that neutral molecules, such as alkanes, *attract* each other mainly through van der Waals forces. Van der Waals forces arises from the rapidly fluctuating dipoles moment (10^{15} sec^{-1}) of a neutral atom, which leads to polarization and consequently to attraction. This is also called the London potential between two atoms in a vacuum and is given as:

$$V_{vdw} = -(L_{11}/R^6) \tag{5.3}$$

where L_{11} is a constant which depends on the polarizability and the energy related to the dispersion frequency, and R is the distance between the two atoms. Since the London interactions with other atoms may be neglected as an approximation, the total interaction for any macroscopic bodies may be estimated by a simple integration.

The electrostatic interactions are known to affect such systems in many important ways. The repulsive electrical potential between any equally charged particles exhibit (φ(distance)):

$$\varphi(distance) \propto 1/distance$$

It is also found that φ(distance) is a long-range force. Accordingly, the presence of ions between these two particles is found to modify the potential and leads to the screening of the interaction as described by Debye–Hückel (**D-H**) theory. When two similarly charged colloid particles under the influence of the **EDL** come close to each other, they will begin to interact. The potentials will feel each other, and this will lead to consequences (Figure 5.5).

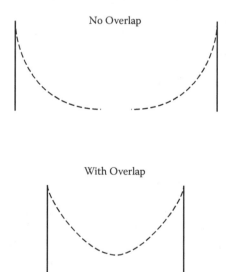

Figure 5.5 Electrostatic interaction between charged particles and magnitude of surface potential (SP).

The charged molecules or particles will be under *both* van der Waals and electrostatic interaction forces. The van der Waals (**vdw**) forces which operate at short distance between particles will give rise to strong attraction forces. The potential of mean force between colloid particles in an electrolyte solution plays a crucial role in the stability and the kinetics of agglomeration in colloidal dispersions.

This kind of investigation is important in various industries:

inorganic materials (ceramics, cements)
foods (milk)
bio-macromolecular systems (proteins and DNA)

The **DLVO** (Derjaguin–Landau–Verwey–Overbeek) theory notes that the stability of a colloidal suspension is mainly dependent on the distance between the particles (Adamson & Gast, 1997; van Oss, 2006; Birdi, 2009). The distance between particles may change depending on various parameters (such as electrolyte concentration; charge on the counter-ion; pH; effect of adsorbed species, such as polymers). **DLVO** theory has been modified in later years, and different versions are found in the current literature.

The electrostatic forces will give rise to repulsion at large distances; see Figure 5.6.

It is known that the electrical charge–charge interactions take place at a large distance of separation. The resultant curve is shown (schematic) in Figure 5.6. The barrier height determines the stability with respect to the quantity **k T**, the kinetic energy. **DLVO** theory predicts, in most simple

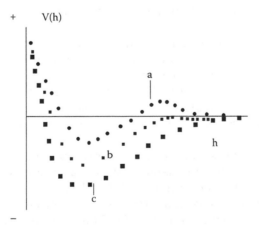

Figure 5.6 Variation of repulsion and attraction forces versus distance between two particles (schematic).

terms, that if the repulsion potential (Figure 5.6) exceeds the attraction potential by a value

$$W \gg k\,T, \tag{5.4}$$

then the suspension will be stable. On the other hand, if

$$W \le k\,T, \tag{5.5}$$

then the suspension will be unstable and it will coagulate.

It has been found, however, that **DLVO** theory does not always provide a comprehensive analysis. It is basically a very useful tool for such analyses of complicated systems. Especially, it is a useful guidance theory in any new application or any industrial development.

In biological cells, adhesion phenomena are known to be critical for function and treatment. These systems have been analyzed in relation to various forces involved in the cell adhesion process (Ruggiero & Mantelli, 2002). The aggregation phenomena and the hydrophobic energy was evaluated by **DLVO** theory.

These examples thus delineate the vast application of **DLVO** theory, whenever one encounters systems where interfacial charges are present. One may need to make some modifications for different systems, but the basic analytical procedure will be based upon the **DLVO** theory or theories (Birdi, 2009).

Charged colloids (electrical charge distribution at interfaces)

The interactions between *two charged bodies* will be dependent on various parameters (e.g., surface charge, electrolyte in the medium, charge distribution) (Figure 5.7).

The distribution of ions in an aqueous medium needs to be investigated in such charged colloidal systems. This observation means that the presence of charges on surfaces means that a potential exists that needs to be investigated. On the other hand, in the case of neutral surfaces, one has only the van der Waals forces to be considered, and hence the added electrolytes show very little or no effect on the physico-chemical properties of the system.

In the case of surfactant (soaps) systems, one finds that the solubility shows a sudden change in characteristics at a certain critical concentration (critical micelle concentration (CMC)). This arises from the fact that surfactants start to form large aggregates, called micelles. This was clearly seen in the case of charged micelles (**SDS** micelles), where the addition of small amounts of NaCl to the solution showed:

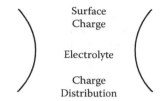

Figure 5.7 Different interaction forces between two charged bodies (schematic).

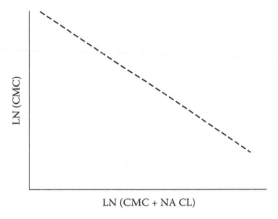

Figure 5.8 Variation of Log (**CMC**) of surfactant solutions as a function of electrolyte concentration (Log (NaCl + CMC); (see text for details) (data for SDS and NaCl solution at 25°C). Note: slope is proportional to surface charge ca. 30%.

*large decrease in **CMC*** in the case of ionic surfactant (Figure 5.8)
almost no effect in nonionic micelles (since in these micelles there are no charges or **EDL**) (Table 5.2)

The surface charges on the micelle are partially neutralized by the counter-ions (sodium ions in the case of SDS micelles). Experiments have shown that if there are 100 monomers per micelle (for example, in SDS micelle, 100 SD⁻ charges), then about 70 Na⁺ ions are bound per micelle. This gives a negative net charge of 30. As one increases the NaCl concentration, more counter-ions are present near the interface of the micelle.
    This can be shown as follows:

```
MICELLE = 100 SD- monomers = 100 negative charges
Number of charges found = 30 negative charges
```

This means that there are approximately 70 Na+ ions adsorbed on the micelle. In fact, this behavior has been found for all the other ionic micelles. It is also obvious that if the charge was zero, then the micelle will

Table 5.2 Variation of Critical Micelle Concentration (CMC) of SDS Micelles with Added NaCl (at 25°C)

Added NaCl	CMC (g/liter)	CMC (mole/liter)
0	2.3	0.008
0.05	0.6	0.0023
0.1	0.3	0.0015
0.2	0.2	0.001
0.3	0.15	0.0006
0.4	0.1	0.0005

be insoluble in water! The observation that the ionic micelles are highly soluble in water indicates that these exhibit a large charge on interface. Furthermore, such specific counter-ion binding is found to be a general behavior for all kinds of systems with charged interfaces:

micelles
macro-ions (proteins, DNA, synthetoc polymers)
solid surfaces (electrodes, battery, colloids)
liquid drops (emulsions, etc.)

However, it is most interesting to note that the change in **CMC** is very sensitive to NaCl concentration. Even less than 0.001 mole/liter of NaCl can be found to decrease the **CMC** appreciably (Table 5.2).

Electrostatic and **EDL** forces are found to play a very important role in a variety of systems as known in science and engineering.

It would be useful to consider a specific example in order to understand these phenomena. Let us take a surface with positive charge, which is suspended in a solution containing positive and negative ions. There will be a definite surface potential, ψ_o, which decreases to a value zero as one moves away into solution (Figure 2.5). The concentration of positive ions will decrease as one approaches the surface of the positively charged surface (charge–charge repulsion). On the other hand, the oppositely charged ions, negative, will be strongly attracted toward the surface. This has been described as the Boltzmann distribution:

$$n^- = n_o \, e^{\,(z\,\varepsilon\varepsilon\,\psi/k\,T)} \tag{5.6}$$

$$n^+ = n_o \, e^{\,-(z\,\varepsilon\varepsilon\,\psi/k\,T)} \tag{5.7}$$

This shows that positive ions are repelled while negative ions are attracted to the positively charged surface. At a reasonably far distance from the particle, $n^+ = n^-$ (as required by the electro-neutrality). Through some simple assumptions, one can obtain an expression for ψ (r), as a function of distance, **r**, from the surface as:

$$\psi \ (r) = z \ e/(D \ r) \ \varepsilon e - \kappa \ r \tag{5.8}$$

where κ is related to the ion atmosphere around any ion. In any aqueous solution when an electrolyte, such as NaCl, is present, it dissociates into positive (Na+) and negative (Cl-) ions. Due to the requirement of electro-neutrality (that is, there must be same positive and negative ions), each ion is surrounded by an oppositely charged ion at some distance. Obviously, this distance will decrease with increasing concentration of the added electrolyte. The expression $1/\kappa$ is called the *Debye length*. It is a measure of the thickness of the diffuse layer and equals the thickness of the equivalent parallel plate condenser.

As expected, the **D-H** theory tells us that ions tend to cluster around the central ion. A fundamental property of the counter-ion distribution is the thickness of the ion atmosphere (Figure 5.9). This thickness is determined by the quantity Debye length or Debye radius ($1/\kappa$). The magnitude of $1/\kappa$ has dimension in cm, as follows:

$$\kappa = (8 \ N \ 2)/(1000 \ k_B \ T) \ \tfrac{1}{2} \ I^{1/2} \tag{5.9}$$

The values of k_B= 1.38 10 $^{-23}$ J/molecule K, **e** = 4.8 10 $^{-10}$ esu. Thus the quantity $k_B \ T \ / \ e$ = 25.7 mV at 25°C.

Using these data one gets for $1/\kappa$:

$$1/\kappa = 3.04 \ 10^{-8} \ (1/I) \ cm$$

$$= 3.04 \ (1/I) \ \text{Å}$$

Debye-Huckel
Charge Zone

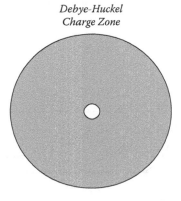

Figure 5.9 The Debye thickness ($1/\kappa$) around a specific ion (schematic).

As an example, with a 1:1 ion (such as NaCl or KBr) with concentration 0.001M, one gets the value of $1/\kappa$ at 25°C (298K):

$$1/\kappa = (78.3\ 1.38\ 10^{-16}\ 298)/(2\ 4\ \Pi\ 6.023\ 10^{17}$$
$$(4.8\ 10^{-10})^2)^{0.5}$$
$$= 9.7\ 10^{-7}\ cm = 97\ 10^{-8}\ cm \tag{5.10}$$
$$= 97\ Å$$

The expression in equation 5.10 can be rewritten as:

$$\psi\ (r) = \psi_o\ (r)\ exp(-\ \kappa\ r) \tag{5.11}$$

which shows the change in $\psi\ (r)$ with the distance between particles (r). At a distance $1/\kappa$ the potential has dropped to ψ_o. This is accepted to correspond with the thickness of the double layer (Table 5.3). This is the important analysis, since the particle–particle interaction is dependent on the change in $\psi\ (r)$. The decrease in $\psi\ (r)$ at the Debye length is different for different ionic strength (Figure 5.10).

Table 5.3 Debye Length ((1/κ) nm) in Aqueous Solutions (25°C)

Salt Concentration	1:1	1:2	2:2 molal
0.0001	30.4	17.6	15.2
0.001	9.6	5.55	4.81
0.01	3.04	1.76	1.52
0.1	0.96	0.55	0.48

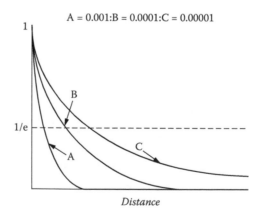

Figure 5.10 Variation (decrease) in electrostatic potential with distance of separation as a function of electrolyte concentration (ionic strength).

The data in Table 5.3 show values of **DH** radius in various salt concentrations. The magnitude of $1/\kappa$ decreases with **I** and with the number of charges on the added salt. This means that the thickness of the ion atmosphere around a reference ion will be much compressed with increasing value of **I** and z_{ion}.

A trivalent ion such as Al^{3+} will *compress* the double layer to a greater extent in comparison with a monovalent ion such as Na^+. Further, inorganic ions can interact with charge surface in one of two distinct ways:

1. nonspecific ion adsorption where these ions have no effect on the iso-electric point
2. specific ion adsorption, which gives rise to change in the value of the iso-electric point

Under those conditions where the magnitude of $1/\kappa$ is very small (e.g., in high electrolyte solution), one can write:

$$\psi = \psi_o \exp -(\kappa\, x) \tag{5.12}$$

where x is the distance from the charged colloid.

The value of ψ_o is found to be 100 mV (in the case of monovalent ions) $(= 4\, k_B\, T/z\, e)$.

Experimental data and theory shows that the variation of ψ is dependent on the concentration and the charge of the ions (Figure 5.11). These data show that:

the surface potential drops to zero at a faster rate if the ion concentration (**C**) increases, and
the surface potential drops faster if the value of **z** goes from 1 to 2 or higher.

In biology, the microbial surfaces are charged particles, and their behavior is dependent on the interfacial charges. A recent analysis on the application of **DLVO** theory on microbial systems was reported (Strevett, 2003). The microbial surface was analyzed regarding hydrophilicity or hydrophobicity, attachment and microbial film attachment.

Colloid phenomena in wastewater technology

Cleaning of wastewater is an enormous challenge to mankind in various technology areas. The wastewater treatment of cellulose–paper industry is one such example. The wastewater in this industry contains mainly biologically nondegradable derivatives of lignin in the form of

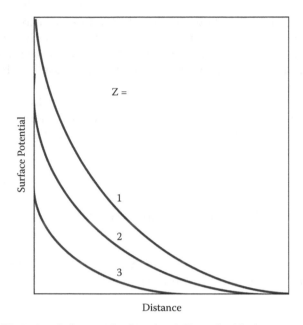

Figure 5.11 Variation (schematic) of in the diffuse double layer as a function of charge on the ions (z).

ligno-sulphonates (LgS). The wastewater was treated with multivalent cations, such as Al+3. These cations neutralize the surface charge of colloid particles of LgS (Sineva, 1991).

Electro-kinetic processes

The charged ions or particles are characterized under two states: *stationary state* and *kinetic state*. More than two centuries ago it became apparent that clay particles, when dispersed in water media, would migrate under the influence of an applied electric field.

```
ELECTRODE (positive)....charged particle.....
ELECTRODE (negative)
```

This observation thus allowed one to estimate the surface charge characteristics. In the following let us consider what happens if the charged particle or surface is under dynamic motion of some kind. The movement of charged species is called the electro-kinetic process. It has been found that movement of charges in solutions produces local asymmetric potentials. Further, there are different systems under which the electro-kinetic phenomena are investigated.

1. **Electrophoresis**: This system refers to the movement of the colloidal particle under an applied electric field. This is the phenomenon of motion of any dispersed particle relative to a fluid under the influence of an electric field. This arises from the fact that particles in general always exhibit a surface charge (positive or negative). When an electric field is applied then through the Coulomb force, the particle will move. This procedure is extensively used in different systems:

   colloidal suspensions (wastewater treatment)
   macro-ions (polymers, proteins)
   biological cells

2. **Electro-osmosis**: In this system a fluid passes next to a charged material. This is actually the complement of electrophoresis. The pressure needed to make the fluid flow is called the *electro-osmotic pressure*.

3. **Streaming potential**: If a fluid is made to flow past a charged surface, then an electric field is created, which is called streaming potential. This system is thus opposite of the electro-osmosis.

4. **Sedimentation potential**: A potential is created when charged particles settle out of a suspension. This gives rise to sedimentation potential, which is the opposite of the streaming potential.

The reason for investigating electro-kinetic properties of a system is to determine the quantity known as the zeta potential.

Electrophoresis, the movement of an electrically charged substance under the influence of an electric field (Figure 5.12), may be related to fundamental electrical properties of the body under study and the ambient electrical conditions by the equation given below. **F** is the force, **q** is the charge carried by the body, **E** is the electric field:

$$F_e = q\ E \tag{5.13}$$

The resulting electrophoretic migration is countered by forces of friction such that the rate of migration is constant in a constant and homogeneous electric field:

$$Ff = v\ f_r \tag{5.14}$$

where **v** is the velocity and f_r is the frictional coefficient.

$$Q\ E = v\ fr \tag{5.15}$$

The electrophoretic mobility μ is defined as follows:

$$\mu = v/E = q/fr \tag{5.16}$$

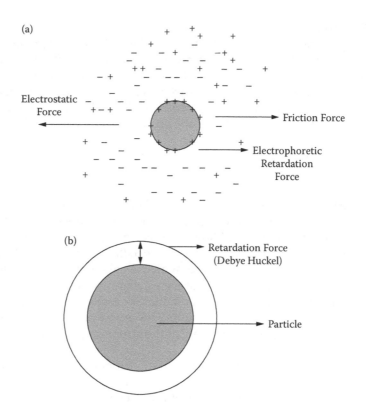

Figure 5.12 Electrophoretic movement of a particle under the influence of an electric field. (a) Friction force—Electrostatic force—Electrophoretic retardation force; (b) Retardation force (Debye–Hückel region).

The expression above applied only to ions at a concentration approaching 0 and in a nonconductive solvent. Polyionic molecules are surrounded by a cloud of counter-ions which alter the effective electric field applied on the ions to be separated. This renders the previous expression a poor approximation of what really happens in an electrophoretic apparatus. The movement is conveniently studied under a microscope as a function of applied field. In some cases one has designed micro-electrophoresis apparatus, where minute amounts of samples can be investigated (Scheludko, 1966).

The mobility depends on both the particle properties (e.g., surface charge density and size) and solution properties (e.g., ionic strength, electric permittivity, and pH). For high ionic strengths, an approximate expression for the electrophoretic mobility, μ_e, is given by the Smoluchowski equation,

$$\mu_e = \varepsilon\, \varepsilon_o\, \eta/n, \tag{5.17}$$

where ε is the dielectric constant of the liquid, εo is the permittivity of free space, η is the viscosity of the liquid, and n is the *zeta potential* (i.e., surface potential) of the particle (or all kinds of charged species). In the case of proteins, one can analyze number of charges on these macro-ions. This is useful in many systems where barely one charge can lead to a significant change in protein function.

Electro-deposition is becoming an important ceramic process-ing technique for a range of applications. The fabrication of electrodes and films for solid oxide fuel cells, fiber reinforced and graded ceramic composites, nano-structured materials, and a range of electronic bio-medical applications are based on electro-deposition and electrophoretic depositions.

In food technology, the charged species are also known to be of impor-tance. The stability and other characteristics of apple juice have been inves-tigated (Benitez et al., 2007). The turbidity of the juice was investigated as a function of pH and ionic strength. The energy barrier between particles was studied as a function of charges on the particles (as determined from **zeta-potential**). The hydration was found to decrease with decrease in pH (and increase in ionic strength).

Electrophoretic **DNA fingerprinting**: Electrophoresis analysis is also being used in the identification of DNA fingerprinting. DNA segments are analyzed by electrophoresis, which are different for different DNA strands (hence the fingerprinting technology application).

Stability of lyophobic suspensions

Particles in all kinds of suspensions or dispersions interact basically with two different kinds of forces (i.e., *attractive forces* and *repulsive forces*). Colloidal dispersions are generally classified into two major classes according to their mode of stabilization:

Lyophilic colloids
Lyophobic colloids

Lyophilic colloids are able to acquire stability by solvation of the interface. The solvation includes all kinds of interactions from mere physical wetting to the formation of adherent thick layers of oriented solvent molecules. Lyophobic colloids are known to be stabilized by an electrostatic repulsion between particles, which is related to the adsorbed ions that are either sorbed onto or dissolved out of the surface of the particle.

The interfacial charges at the surface need to be analyzed by the understanding of the ions in the interface and surrounding. The ions at interfaces exhibit double layer. The specific ions are adsorbed at the surface of the particles, thus giving rise to net charge. In some cases, the charge can arise from specific ions, originally present in the surface of the particle. The charged surface attracts ions of the opposite sign toward it, but they are retained in the medium due to the thermal (kinetic) energy $(k\ T)$. This thus leads to the creation of an electrical double layer in the vicinity of each particle.

One observes that lyophobic suspensions (sols) must exhibit a maximum in repulsion energy in order to have a stable system. The total interaction energy, V (h), is given as (Scheludko, 1966; Bockris et al., 1980; Adamson & Gast, 1997; Chattoraj & Birdi, 1984; Birdi, 2002; 2009):

$$V(h) = V_{el} + V_{vdw} \tag{5.18}$$

where V_{el} and V_{vdw} are *electrostatic repulsion* and van der Waals *attraction* components. Dependence of the interaction energy $V(h)$ on the distance h between particles has been ascribed to coagulation rates as follows:

1. During slow coagulation
2. When fast coagulation sets in

The dependence of energy on h and V(h),

$$V(h) = [(64\ C\ RT\ \psi^2)/k\ exp(-k\ h) - H/(2\ h^2)] \tag{5.19}$$

satisfies the requirements of this coagulation rate. For a certain ratio of constants it has the shape shown in Figure 5.6. For large values of h, $V(h)$ is negative (attraction), following the energy of attraction V_{vdw}, which decreases more slowly with increasing distance $(\sim 1/h^2)$. At short distances (small h), the positive component Vel (repulsion), which increases exponentially with decreasing h, $(exp(-k\ h)$, can overcompensate V_{vdw} and reverse the sign of both $dV(h)/dh$ and $V(h)$ in the direction of repulsion. On further reduction of the gap (very small h), V_{vdw} should again predominate, since:

$$V_{el} = 64\ C\ RT\ \psi^2/k, as\ h \to 0, \tag{5.20}$$

whereas the magnitude of V_{vdw} increases indefinitely when h > 0. There is thus a repulsion maximum in the function $V(h)$, which can be easily found from the condition $dV(h)/dh = 0$. The choice of solution (maximum or minimum) does not present any difficulty since V(h) is positive for the maximum. From these relations, one finds that when the electrolyte concentration is increased, the magnitude of k in the exponent of V_{el} also increases (compression of diffuse layers), so that the maximum caused by it becomes lower. At a certain value of c the curve $V(h)$ will become similar to curve b in Figure 5.6 with $V_{max} = 0$. Accordingly, the coagulation process will become fast starting from this concentration. This corresponds to the *critical concentration*, C_{cc}. In other words, the magnitude of critical concentration (**cc**) can be estimated from simultaneous solution of the following:

$$dV(h)/dh = 0 \text{ and } V(h) = 0. \tag{5.21}$$

One can write the following:

$$dV(h)/dh = [-(64 \ C_{cc} \ \mathbf{RT} \ \psi^2)/k \ \exp(-k_{cr} \ h_{cr}) + \mathbf{K}/(h_{cr}{}^3)] = 0 \tag{5.22}$$

and

$$V(h) = [(64 \ C_{cc} \ \mathbf{RT} \ \psi^2)/k_{cr} \ \exp(-k_{cr} \ h_{cr}) - \mathbf{K}/(2 \ h_{cr}{}^2)] = 0. \tag{5.23}$$

After expanding these expressions, as related to h and C, this becomes (**Schultze–Hardy Rule**) (for suspensions in water)

$$C_{cc} = 8.7 \ 10^{-39}/Z^6 \ \mathbf{A}^2$$
$$C_{cc} \ Z^6 = constant, \tag{5.24}$$

where the term *constant* includes A (Hamaker constant = approximately $4.2 \ 10^{-19}$ J). The magnitudes of the critical concentration of ion⁻ to the *sixth power* of various ions are inversely proportion to the valency (Z):

$$\mathbf{Z} = 1{:}(2^6) \ 0.016{:}(3^6) \ 0.0014{:}(4^6) \ 0.000244. \tag{5.25}$$

This shows that addition of indifferent ions (which have no specific interactions with the interfacial charges) can nevertheless affect the behavior of the colloid system. This arises from the contraction of the **EDL** region.

The flocculation concentrations of mono-, di-, and trivalent (etc.) gegen-ions should from this theory be expected as:

$$1 : (½)^6 : (1/3)^6{:}.....$$

It thus becomes obvious that the colloidal stability of charged particles is dependent on

1. *concentration of electrolyte*
2. *charge on the ions*
3. *size and shape of colloids*
4. *viscosity of the medium*

The critical concentration (*critical coagulation concentration*) is thus found to depend on the type of electrolyte used as well as on the valency of the counter-ion. It is seen that divalent ions are 60 times (2^6 factor) as effective as monovalent ions. Trivalent ions are several hundred times more effective than monovalent ions. However, ions which specifically adsorb (such as surfactants) will exhibit different behavior. In all washing processes, the composition of the detergents always contains polyvalent electrolytes (such as poly-phosphates, etc).

It is important to mention that this result shows clearly that the **EDL** theory is basically needed to understand these systems. This also clearly shows that the electrical charges at interfaces are radically different from those found in the bulk phases.

Based on these observations, in washing powders composition, one has used multivalent phosphates, for instance, to keep the charged dirt particles from attaching to the fabrics after having been removed (see *Scultze–Hardy rule*). Another example is the wastewater treatment, where for coagulation purposes one uses multivalent ions.

The wastewater is generally treated by sedimentation followed by a biological selector and followed by a solid retention process. The pH is adjusted accordingly (typical values are 6.8 –7.4). In a laboratory one takes samples and determines the necessary ionic strength for optimum sedimentation conditions. Mixtures of KCl and $CaCl_2$ are used (range of concentrations: 0.00001–0.07 M).

Direct interaction force measurement

It is obvious that the direct force measurements at interfaces would be much more useful for determining the behavior of real systems. Recently, scanning probe microscopes (**SPM**) have been used to investigate such systems (Birdi, 2003). The principle in SPM is that a sensor is made to scan over any surface under controlled conditions. The interaction between the sensor and the substrate is monitored (Figure 5.13).

Atomic force microscope (**AFM**) is based upon where the sensor is related to the force acting between the surfaces. The magnitude of surface forces at a SiO_2 particle was investigated by using **AFM** (Lin et al., 1993). In such a system

```
SiO₂....water
```

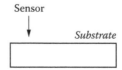

Figure 5.13 Principle of SPM (see text for details).

the electrical surface potential ψ_{oxide} of oxide material in the aqueous media is related to the charges. The Si–OH groups on the surface will form Si–OH_{2+} and SiO– charge sites. The point-of-zero charge (**PZC**) is given as:

$$\mathbf{PZC} = (pK^+ + pK^-)/2$$

where pK^+ and pK^- are defined for the site acting as a proton donor or proton acceptor, respectively.

According to the **DLVO** theory the electrostatic force between two surfaces (such as a sphere and a plane) is given as $\mathbf{F}_{dlvo}$:

$$\mathbf{F}_{dlvo} = (\mathbf{R}_{radius}) (\psi_{sphere} \psi_{plane}) \exp(-d/k)/k$$

where $\mathbf{R}_{radius}$ is the radius of the sphere, the surface potentials of sphere are (ψ_{sphere}) and plane (ψ_{plane}), distance of separation is **d**, and κ is the **D-H**.

It is important to notice that the magnitude of $\mathbf{F}_{dlvo}$ is dependent on:

distance of separation
D-H length
surface charges

The change in sign of p will thus make the F_{dlvo} change from attractive (when p are of opposite signs) to repulsive (when sign of p is the same). This can be achieved in this system by changing the pH.

In the **AFM** apparatus one can measure the force between the tip and the sphere as a function of distance (Birdi, 2003; Lin et al., 1993):

```
AFM Tip (Si3N4) . . . . . . . . . . . . . . . WATER . . . . . . . . . . Substrate
```

The **PZC** for the tip was estimated from the *force* versus *distance* (less than micrometer) curves as a function of pH. This procedure is almost a standard application in all commercial **AFM**. The value of **PZC** was found to be 6.0, which agreed with the data from other methods.

The surface forces between alumina colloid and substrate (sapphire) have been investigated by using AFM method (Polat et al., 2007). The charges were changed from negative–*zero*–positive by pH. The force curves showed a repulsive barrier at a separation of 10 nm (8 nN force). An attractive minimum region was observed at a distance of 2 nm (30 nN force), at acidic pH. Attractive force was present with a deep minimum at 2 nm (45 nN force). These data agreed very satisfactorily with the **DLVO** theory.

As it is obvious, it is not a simple matter to measure the forces between two particles at very short distances. Therefore, one finds different techniques in achieving this information. In a recent study the movement of particles near surfaces was studied by using the total internal reflectance microscopy (**TIRM**) (Flicker et al., 1993). This method allows one to measure the Brownian movement. The resolution is approximately 1 nm. The scattered light is dependent on the orientation of the particle. **TIRM** studies were carried on a system based on

glass slide....*polystyrene latex sphere* (diameter range: 7 to 15 μm).

The double-layer interaction was varied by changing the ionic strength (I) from 0.2 to 3.0 mM. For these systems, the double-layer potential energy, VR(d), between a spherical particle and a flat plate was based on the following equation:

$$VR(d) = 16 \ e \ \mathbf{R} \ (kT/z \ e) \ tanh \ (z \ e \ \varphi_s/4 \ \mathbf{k} \ \mathbf{T})$$

$$Tanh \ (z \ e \ \varphi_p/4 \ \mathbf{k} \ \mathbf{T}) \ exp \ (-\mathbf{k} \ \mathbf{d})$$

where **e** is the dielectric constant of the fluid, **R** is the radius of the sphere (polystyrene), **kT** is the thermal energy, **z** is the ion valency, e is the protonic charge, φ_s and φ_p are the surface potentials of the sphere and plate, respectively. In this model **VR(R)** depends exponentially on separation distance, R. The gravity force, **VR(gravity)**, was as follows:

$$VR(gravity) = 4/3 \ \pi \ \mathbf{R}^3 \ \Delta\rho \ \mathbf{g} \ \mathbf{R}$$

Δ**d** is the difference in density between the particle and the fluid. The magnitude of φp was set equal to −60 to −80 mV, as related to the data from the zeta potential of silica particles. Using this value, the φ_s of polystyrene was estimated as −15 to −30 mV. These data were acceptable and showed that the **DLVO** theory gives an acceptable estimate.

In a recent study (Missana and Adell, 2000) the stability of Na-montmorillonite colloids was investigated. In this study the surface charge behavior was investigated by using dynamic light scattering. The data were analyzed with the help of modified **DLVO** theory.

The surface forces at silica were investigated directly (Adler et al., 2001). These studies showed that the surface has a silica–gel layer, which has a strong effect on the **DLVO** forces. These investigations are of much importance for various ceramic industrial products.

Streaming potential: The interface of a mineral (rock) in contact with aqueous phase exhibits surface charge. The currently accepted model of this interface is the **EDL** model of Stern. Chemical reactions take place between the minerals and the electrolytes in the aqueous phase, which results in net charge on the mineral. Water and electrolytes bound to the rock surface constitute the Stern (or Helmholtz) layer. In this region the ions are tightly bound to the mineral, while away from this layer (the so-called electrical double layer, **EDL**), the ions are free to move about. Further, all ions in water are found to move about along with a given number of solvent molecules (varying from 5 to 10 molecules of water). Since the distribution of ions (positive and negative) is even in the diffuse region, there is no net charge. On the other hand, in the Stern layer there will be asymmetric charge distribution, such that one will measure from zeta-potential data that the mineral exhibits a net charge.

References

Adamson, A. and Gast, E., *Physical Chemistry of Surfaces*, Wiley, New York, 1997.
Adler, J.J., Rabinovich, Y.I, and Moudgil, B.M., Adhesion between nanoscale rough surfaces, *J. Colloid Interface Sci.*, 237, 249, 2001.
Benitez, E.I., Genovese, D.B., and Lozano, J.E., *Food Hydrocolloids*, 21, 100, 2007.
Birdi, K.S., *Scanning Probe Microscopes*, CRC Press, Boca Raton, FL, 2003.
Birdi, K.S., Ed., *Handbook of Surface and Colloid Chemistry*, CRC Press, Boca Raton, FL, 2002; 2009.
Bockris, J.O.M., Conway, B.E., and Yeager, E., Eds., *Comprehensive Treatise of Electrochemistry*, Vol. 1, New York, 1980.
Chattoraj, D.K. and Birdi, K.S., *Adsorption and the Gibbs Surface Excess*, Plenum Press, New York, 1984.
Flicker, S.G., Tipa, J.L., and Bike, S.G., Repulsion between a colloid sphere and a glass plate, *J. Colloid Interface Sci.*, 158, 317, 1993.
Lin, X.Y., Creuzet, F., and Arribart, H., Atomic force microscopy of surfaces, *J. Phys. Chem.*, 7272, 97, 1993.
Missana, T. and Adell, A., On the applicability of DLVO theory, *J. Colloid Interface Sci.*, 230, 150, 2000.
Polat, M., Sato, K., Nagaoka, T., and Watari, K., Normal interactive forces (online), 3, 1883, 2007.
Ruggiero, C., and Mantelli, M., Molecular, cellular and tissue engineering, *Proceedings of the IEEE-EMBS*, 136, 2002.
Scheludko, A., *Colloid Chemistry*, Elsevier, New York, 1966.
Sineva, A.V., Adsorption and ion-exchange, *Colloid Journal*, 53, 740, 1991.
Strevett, K.A. and Chen, G., Microbial surface thermodynamics, *Research in Microbiology*, 154, 329, 2003.

Appendix

General electrochemistry

The chemistry of all kinds of systems is related to different properties of the materials. One subject, *electrochemistry*, concerns the systems involving ions (electrons) and conduction.

In physics one learns that a hydrogen atom is neutral because it consists of one positively charged nucleus and a proton surrounded by a negative charged particle, an electron (Figure **A**.1). The mass of a proton is approximately 2000 times larger than that of an electron.

In uranium (U), a much larger atom, one finds there are 92 electrons surrounding the nucleus consisting of 92 protons and 148 neutrons (a neutron is a neutral particle composed of one proton and one electron).

By definition, an *ion* is an atom or molecule that has *lost or gained* one or more *electrons*. This state gives rise to a positive or negative electrical charge property. *Anion* is a negatively charged species (such as Cl^- or SO_4^-). A *cation* is a positively charged species (such as H^+).

In general, objects may possess a property known as *electric charge*.

If an electrical field is present, then it exerts a force on the charged object. The charged object will accelerate in the direction of the force, in either the same or the opposite direction. Classical mechanics explores these concepts such as force, energy, potential, and so forth in much more detail. Force and potential are directly related. The potential energy decreases as an object moves in the direction that the force accelerates. An example of this is the gravitational potential energy of a stone at the top of a hill is greater than that at the base of the hill. In other words, the potential energy decreases as the object falls, and the energy is translated to motion, or kinetic energy.

However, in *solutions* the ions are much larger and the surroundings media (water) imparts many special properties. Na^+ ion is approximately 35 times larger than H^+. The presence of charged ions gives rise to various characteristics. The charged ions as found when an electrolyte is dissolved in water (such as NaCl) give rise to many important physicochemical properties. If one places two platinum electrodes in this solution, one finds that the conductivity of the solution increases as the concentration of NaCl increases.

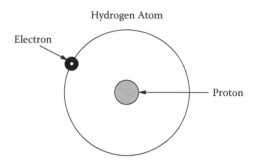

Figure A.1 The hydrogen atom (electron and proton).

In other systems, one may pass current and there may be deposition of cation on the electrode (as described in lead storage). In the latter case one finds that the current needed to deposit a molar quantity is the same for different cations (although related to the number of charges, **z**, on the cation).

    Faradays Law: If current is passed in a system with an electrode and an electrolyte solution, then a given amount of substance is deposited. This is described by the Faradays law as follows. If a current of magnitude **I** (ampere) is passed through an electrolyte solution for **t** seconds, then **I** t coulombs pass through it. The amount, **gm**, which is deposited on the electrode is given as:

$$\mathbf{gm} = (\mathbf{M/z})/(96{,}493)\,(\mathbf{I\ t})$$

where **M/z** (weight of gram molecule divided by the valence), 96,493 is the Faradays charge (i.e., the charge with which 1 gram equivalent of a substance is deposited). One can thus write:

$$\mathbf{I} = (96{,}493\ \mathbf{g\ z})/(\mathbf{M\ t})$$

The ampere, I, was historically a derived unit, defined as 1 coulomb (**C**) per second. One **C** is the amount of electric charge transported in 1 second by a steady current of 1 Ampere (**A**). It is also known that positive and negative charges are usually balanced out. According to Coulombs law, two point charges of +1 **C**, one meter apart, would experience a repulsive force of 9 10^9 N (approximately equal to 900,000 tons weight!).

    In one experiment by using the data as follows one can estimate I:

Hydrogen evolution

time = 1 hour
Hydrogen evolved (standard temp and pressure) = 87 cc
Amount (gram) of H_2 = (87 × 2) / 22,415
                       = 7.82 10^{-3} g
                       = 7.82 mg

Thus:

$$I = (96493 \times 7.82 \ 10^{-3})/(1 \times 3600)$$
$$= 0.208 \ \mathbf{A}$$

Silver deposition

time = 1 hour
Ag deposited = 0.845 g

Thus:

$$I = (96493 \times 0.845) / (107.88 \times 3600)$$
$$= 0.210 \ \mathbf{A}$$

This shows a good agreement between theory and experiment.

It is thus clear that a balance exists between current production and mass balance. In other words, with systems as storage battery, these considerations are important.

Reversible electrodes

It is of interest to consider briefly the electrochemistry of reversible electrode systems. A reversible electrode is one in which each ion of each phase contains a common ion that is free to cross the interface. This may be as:

```
Ag electrode in contact with a AgI solution
```

The Ag^+ and I^- ions will compete for adsorption on the surface of Ag electrode. The situation is described by Nernst equation. It relates the surface potential, ψ_o, to the ratio of concentration of ions with respect to solubility product. If one considers a system of Ag in AgI solution in water as follows:

saturated solution of AgI in water = 8.7 10^{-9} mole/liter
solubility product of AgI = 7.5 10^{-17} mole/liter

The Nernst equation gives the magnitude of ψ_o. So:

$$\psi_o = \mathbf{k} \ \mathbf{T} \ / \ \mathbf{e} \ln (\mathbf{C} \ / \ \mathbf{C}_{zp})$$

where $\mathbf{C}$ is the concentration of Ag+, and $\mathbf{C}_{zp}$ is the concentration of Ag+ at which the AgI exhibits zero charge. In this system we have:

$$\mathbf{C} = 8.7 \ 10^{-9} \ \text{mole/liter}$$
$$\mathbf{C}_{zp} = 3 \ 10^{-6} \ \text{mole/liter}$$

From these data one gets for ψ_o:

$$\psi_o = 2.303 \; \mathbf{R} \; \mathbf{T/F} \log (\mathbf{C/C_{zp}})$$
$$= 25.7 \ln (8.7 \; 10^{-9}/3 \; 10^{-6})$$
$$= -150 \; mV$$

This is a useful example to understand the potentials and surface charges as discussed throughout this book.

Conductance in electrolyte solutions

Pure water exhibits very high resistivity because it has no ions. On the other hand, if ions are present in aqueous media, then the conductivity increases. The conductivity of an aqueous electrolyte solution is measured by using a cell consisting of two Pt electrodes placed at a distance of 1 cm. The cell constant is determined from known solutions as used or calibration. A typical cell consists of the following:

```
Pt........solution........Pt
```

The resistance of the solution is measured by connecting the two Pt electrodes (generally of 1 cm^2 surface area at separation of 1 cm), by using a suitable resistance meter. This can be described as follows.

A solution of KCl, 0.01 N, shows a resistance of 1748.56 Ω. In the same cell a 0.001 M solution of $AgNO_3$ had a resistance of $1.88974 \; 10^4 \; \Omega$.

The specific conductivity of an electrolyte solution, $\mathbf{K}$, is known to be inversely proportional to its resistance, $\mathbf{R}$:

$$\mathbf{K} = k_{cell}/\mathbf{R}$$

where k_{cell} is the cell constant. The value of the latter can be estimated as follows. From the data of KCl solution we can obtain:
$\mathbf{K}$ of KCl solution = $1.41145 \; 10^{-3} \; \Omega^{-1} \; cm^{-1}$ (25 C)

$$k_{cell} = \mathbf{K} \; \mathbf{R}$$
$$= 1.41145 \; 10^{-3} \times 1748.56$$
$$= 2.468 \; cm^{-1}$$

The data of Ag NO_3 solution can be analyzed as follows:
$AgNO_3...0.001$ M

$$\mathbf{K} = 2.468 / 1.88974 \; 10^4$$
$$= 1.306 \; 10^{-4} \; \Omega^{-1} \; cm^{-1}$$

Table A.1 Potassium Bromate Aqueous Solutions

Concentration (gram eq liter⁻¹)1000	K/1000	Λc
147	15.3000	103.62
93	10.067	107.53
49	5.496	112.26
11	1.332	120.34
3.28	0.4076	124.18
0.5443	0.0693	127.32

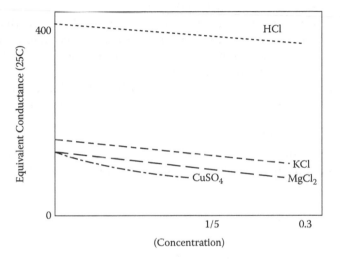

Figure A.2 Plots of conductivity versus concentration of various electrolytes in water.

The data in Table A.1 has been found from experimental measurements for potassium bromate, $KBrO_3$, aqueous solutions. One notices that the magnitude of **K** is fairly linear dependent on concentration of the electrolyte. The specific, **K**, and equivalent, Λc, conductivities are defined as:

$$\Lambda c = 1000\ \mathbf{K}/\ (\mathbf{C})$$

where **C** is in equivalent liter⁻¹ units. It is noticed that the magnitude of Λc decreases with increasing concentration. The simple reason one can deduce is that as the distance between ions in the medium decreases, the ion–ion interactions become greater and thus nonideal behavior is expected (Figure A.2).

The plots of Λc versus concentration are used to estimate equivalent conductivity at infinite dilution, Λo. The data in Table A.1 provide a value of Λo equal to 129.3 M⁻¹ cm².

Electrical double layer (EDL)

Adsorption of ions at interfaces

The charge distribution (that is, the distances between positive and negative ions) in a solution of electrolytes is perturbed if a charged metal electrode is inserted into this media. A solid electrode (for example a metal: Zn) in contact with an aqueous solution of an electrolyte (NaCl, etc.) generally carries a surface charge. This kind of situation is common in almost all kinds of storage batteries. At the surface of the electrode there will thus be a definite potential, ψ_o. The magnitude of ψ_o will decrease as one moves away from the surface (Figure 2.5).

The surface potential decreases close to the interface due to the strongly adsorbed counter-ions (Stern layer). Thus the magnitude of ψ_x will be dependent on distance, x. The nature of ψ_x and the distribution of ions in such a system will be analyzed here.

In literature one finds various models which describe this situation of charges. The most fundamental are based on the following analyses.

Helmholtz model

According to this model, a positively charged electrode will interact with equal number of negatively charged ions, within the so-called Helmholtz plane.

```
POSITIVE ELECTRODE
++++++++++++++++++++
--------------------......HELMHOLTZ
```

This means that the magnitude of ϕ_o will rapidly decrease to zero with x. This model has been found to be unreasonable, since it does not correspond with experimental data. Further, this simple model does not consider the movement of ions in the solution as well as the adsorption of solvent on the electrode. Accordingly, one finds that other modified models of the surface potential were analyzed.

Gouy–Chapman model

The most plausible model would be to expect the counter-ions (negative ions) to be strongly attracted toward the electrode with a positive charge.

```
POSITIVE ELECTRODE
+++++++++++++++++++++++++++++++++
-+----+----+-----+-----+----- ......GOUY-CHAPMAN
-+-+-+----+-+-+-+------------
-+-+-+-+-+-+-+-+-+-+-+-+-+-
```

The different charge density states can be explained as follows. The fixed charge density is denoted as ψ_o, the Stern layer consists of ψ_{st}, and the Gouy layer consists of ψ_{go}. In a system with electroneutral state we have:

$$\psi_o + \psi_{st} + \psi_{go} = 0 \tag{A.1}$$

It is thus obvious that the state of charge distribution is strongly dependent on the Debye–Hückel thickness, $1/\kappa$.

The distribution of ions will be dependent on the distance, x, and the potential ψ_x. Using the Boltzmann distribution theory under such potential considerations, one finds the following relations. The number of ions per unit volume, N, of different charges will be given as:

$$\text{Positive ions: } N^+ = N^+_{bulk} \exp\left(- z^+ \, e \, \psi/k_B \, T\right) \tag{A.2}$$

$$\text{Negative ions: } N^- = N^-_{bulk} \exp\left(+ z^- \, e \, \psi/k_B \, T\right) \tag{A.3}$$

where $N_{i,bulk}$ is the number of ions of i ions in unit volume, z_i is the charge number, e is the electronic charge, k_B is the Boltzmann constant (=), and **T** is the temperature.

It is thus seen that near the positive electrode there will be some degree of imbalance of electrical charges. There will be more negative charge than positive very close to the electrode. This clearly will have some specific effects on the system (e.g., in the case of a storage battery).

The quantity charge density, ρ, is defined as:

$$\rho = \Sigma \, (Ni \, z_i \, e) \tag{A.4}$$

$$= \Sigma \, (N_{i,bulk} \, zi \, e \exp\left(-zi \, e \, \psi/k_B \, T\right)) \tag{A.5}$$

In the case for a symmetrical electrolyte (NaCl), $Z+ = z- = z$), one can derive the following:

$$\rho = 2 \, N_{bulk} \, z \, e \, Sinh \left(- z \, e \, \psi/k_B \, T\right) \tag{A.6}$$

From the well-known Poisson equation for a plane interface one has the following relation:

$$d^2 \, \psi/dx^2 = - \rho/\varepsilon \tag{A.7}$$

where ε is the permittivity of the liquid. Combining these last three equations one can obtain:

$$d^2 \, \psi/dx^2 = - \Sigma \, (N_{i,bulk} \, zi \, e/ep) \, Exp \, (zi \, e \, \psi/k_B \, T)) \tag{A.8}$$

and for a symmetrical electrolyte :

$$d^2 \psi/dx^2 = -2 \, (N_{bulk} \, z \, e/\epsilon) \, Sinh \, (- z \, e \, \psi/k_B \, T)) \qquad (A.9)$$

By using the limits in the expansion of this relation arising from the electroneutrality condition, one obtains:

$$d^2 \psi/dx^2 = \kappa \, \psi \qquad (A.10)$$

where κ is the Debye–Hückel length, defined as:

$$\kappa^2 = (2 \, e2 \, NA/\epsilon \, k_B \, T)I \qquad (A.11)$$

where I (mol m^{-3}) is the ionic strength:

$$I = 1/2 \, NA \, \Sigma_i \, (N_i \, zi^2)$$

In aqueous solutions, for 1:1 electrolyte, the Debye length is found to be:

$$1/\kappa = 3/c^{1/2} \, (\text{Å}) \qquad (A.12)$$

The quantity $1/\kappa$ has the dimensions of length. This corresponds to the region where the electrical double layer exists. One notices that the magnitude of 1/k decreases as **I** increases.

This means that as two charged particles approach each other, the degree of interaction at a given distance x will depend on the ψx, which in turn depends on I. This can be depicted as follows in the case of two particles with the same charge:

Large $1/\kappa$...........Interaction (repulsion) at large distance
Small $1/\kappa$..........Interaction (repulsion) at small distance

These observations have been found to be useful in many systems, such as:

colloidal systems
electrodes (battery technology)
biology (electrophysiology)
adhesion and friction

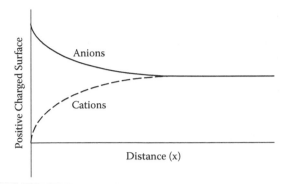

Figure A.3 Variation of cations and anions versus distance from the charged interface (schematic).

microfluid technology
microsensors

It is thus useful to analyze the state of potential in a real system as a function of distance, x, and the above equations. The variation of ψ can be derived from equation A.10:

$$\psi_x = \psi_o \text{ Exp} (-\mathbf{k} \, x) \tag{A.13}$$

This means that ψ_o decreases with x. Due to the positive charge on the solid, a greater number of negative ions will be concentrated near its surface, as compared with positive ions. The value of ψ_o will thus decrease with x. This indicates the state of potential as one moves away from the surface toward the bulk of solution. The value of ψ_x reaches zero at a certain distance, which has been identified with $1/\kappa$. The variation of positive and negative ions is depicted in Figure A.3 GUCHap3.

Applications of double layer phenomena in industry

It must be obvious from the above that the number of systems where charged ions are involved in everyday life systems must be very large. Since charged surfaces (as solid or liquid drops) are found in a variety of systems in everyday life, some important examples are given in the following sections. This is especially true due to the greater understanding of electrical phenomena in recent decades, which has made a major impact.

Colloidal systems

The *colloidal dispersions* as used in various industrial products are analyzed with regard to:

> formation
> stability
> instability
> utilization and exploitation
> handling

Everyone is aware of inkjet printer technology. Mankind has used ink for over 3000 years! This system is mainly based on colloidal technology. Most simple ink was made of finely powdered charcoal and olive oil. This requires ink jets to be produced at a great accuracy regarding both flow rates and quantity per jet.

The energy sector is becoming a much greater concern as one needs to meet the new challenges (pollution control and efficiency).

Industrial applications

Batteries: different storage battery

Technological advances as found in battery development are some of the most important areas where interfacial charge has played an important role. One of the most basic inventions affecting the everyday life of mankind is the battery.

Battery technology is very complex and extensively investigated. In all battery technology devices, the basic principle is based on interfacial charges as depicted below:

ANODE SURFACE CHARGE: : : : *ELECTROLYTE*: : : :
CATHODE SURFACE CHARGE

In other words, the battery charge or discharge processes take place at the interface of the anode and cathode. A battery is composed of several electrochemical cells that are connected in series or parallel to achieve the necessary voltage. Each cell is primarily composed of an electrode in contact with an electrolyte media (Figure A.4). The interfacial charges are the basic chemical process.

This is obvious when considering that a battery is very essential in everyday life. A complete description of this technology is thus out of the scope of this book. However, a short introductory analysis will be presented here, along with some important examples. For instance, every car is started with the help of a battery. All laptop computers and mobile

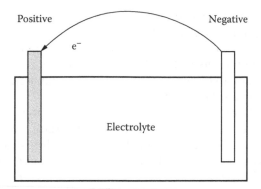

Figure A.4 A typical battery cell (composed of two electrodes in contact with electrolyte media).

LEAD ACID BATTERY
POSITIVE----PbO_2/$PbSO_4$
$H+$/H_2O/SO_4^-
(H_2SO_4/H_2O)
NEGATIVE----Pb/$PbSO_4$

Figure A.5 Lead–acid battery structure and reactions.

phones are exclusively run on rechargeable batteries. In the latter case the battery can be recharged for about 1000 cycles. Many complex instruments (such as an artificial pacemaker) are dependent on a battery. The aim of a battery is to provide electrical current. In other applications the battery is a very useful storage device of energy (such as wind or wave energy).

A typical lead–acid battery (as used in car starter) is made up of lead electrodes and H_2SO_4 (Figure A.5). The reactions that take place are depicted; however, the exact chemical interfacial reactions are more complex.

A mercury battery (also called mercuric oxide battery, or mercury cell) is a nonrechargeable electrochemical battery, a primary cell. Due to the content of mercury and the resulting environmental concerns, the sale of mercury batteries is banned in many countries.

Sodium hydroxide or potassium hydroxide are used as an electrolyte. Sodium hydroxide cells have nearly constant voltage at low discharge currents, making them ideal for hearing aids, calculators, and electronic watches. Potassium hydroxide cells, in turn, provide constant voltage at higher currents, making them suitable for applications requiring current surges, for example, photographic cameras with flash, and watches with a backlight. Potassium hydroxide cells also have better performance at lower temperatures. Mercury cells have very long shelf life, up to 10 years.

Mercury batteries use either pure mercuric oxide or a mix of mercuric oxide with manganese dioxide as the cathode. The anode is made of zinc and separated from the cathode with a layer of paper or other porous material soaked with electrolyte. During discharge, zinc oxidizes to zinc oxide and mercuric oxide gets reduced to elementary mercury. Mercury batteries are very similar to silver-oxide batteries.

Mercury batteries using mercury(II) oxide cathode have a very flat discharge curve, holding constant 1.35 V (open circuit) voltage until about the last 5% of their lifetime, when their voltage drops rapidly. Mercury batteries with cathodes made of a mix of mercuric oxide and manganese dioxide have output voltage of 1.4 V and more sloped discharge curve. A more recent type of battery is based on Li ion reactions (Figure A.6). The reactions involved are rather complex.

A simple comparison of different batteries shows that the energy density characteristics are variable (Table A.2). Many technical problems arose when different kinds of restrictions were imposed, such as banning the sale of mercury oxide batteries, which caused numerous problems for photography technology because a great amount of equipment frequently relied on the advantageous discharge curves and long lifetime of these batteries.

Alternatives used are zinc–air batteries, with a similar discharge curve but much shorter lifetime (a few months) and poor performance in dry climates; alkaline batteries, with voltage widely varying through their lifetime; and silver–oxide batteries, with higher voltage (1.55 V) and a very flat discharge curve, making them possibly the best, though expensive, replacement.

LITHIUM ION BATTERY

(3.8 volt)

POSITIVE----$Li(x)CoO_2/Li\,Co\,O_2$

$Li+$

$LiPF6$

NEGATIVE----$LiC6/C6$

Figure A.6 Lithium (Li) ion battery structure and reactions.

Table A.2 Battery Comparison

Battery	Energy density	
	Watts hr/liter	Watts hr/kg
lead acid	80	25
Ni–Cd	150	50
Li ion	300	150

Metal corrosion phenomena (an interfacial charge phenomena)

The enormous cost involving the corrosion process is well known. Therefore, a large amount of current literature exists in reference to corrosion science. If a metal surface is exposed to a salt solution, the following may be observed after some time:

The metal surface is unchanged, such as a gold plate in water.
The metal surface is attacked by the solution as found from deterioration of the metal (such as an iron plate in water). The metal will be unstable so long as the process

$$metal = metal(z+) + z \ e-,$$

where the electrons are retained by the metal, proceeds spontaneously. The *corrosion* process is related to the potential and currents involved in such surface phenomena. Metals such as iron, zinc, aluminum, and others are called *base* metals. These metals transform into ionic salts or oxides with a release of free energy. In one study it was found that the rate of corrosion of Al in HCl (20%) was related to the purity of Al, Table A.3.

It is a general phenomena that corrosion rates of pure metals are lower than those for impure metals. This arises from the fact the interfacial charges are different for impure components. Corrosion can also be decreased by using different methods (paints, inhibitors). Recently it has been found that by adding very small amounts of suitable surface active agents, one can decrease the corrosion appreciably (due to the surface adsorption of the surface active agents). This has been applied to storage battery technology.

It is generally accepted that the corrosion process is the deterioration of a metal structure initiated from its surface. In other words, if the surface could be treated in such a way as to retard or inhibit this process at the surface, then one can avoid the corrosion.

Table A.3 Rate of Dissolution of Al in HCl Solution

Purity of Al	Loss of weight (g/ml/day)
99.998	6
99.990	112
99.970	6500
99.880	36,000
99.200	190,000

Biological sensors

Sensors in biology

In recent decades, a variety of sensors have been commercially made available for detection and analyses of biological molecules in blood or urine and so forth. In fact, this trend is going to be more and more common for a variety of biological tests. The determination and control of glucose in blood is an important application for diabetic patients. The microsensor is based on the selective reaction between the glucose and enzyme (glucose oxidase):

```
                        ENZYME
      glucose + O2 ------------> H2O2 + gluconic acid
```

The electrode is calibrated with standard solutions (varying from 2 to 40 mmol glucose/liter). The sensitivity (using only a tiny drop of blood (less than 10 µL)) and reliability has been found to be very high for diabetic control.

Lipid monolayers—interfacial charges

As is apparent from the above explanation, the surface property of water phase plays an important role in everyday phenomena. The characteristics of the surface of water changes appreciably if a monolayer of lipid is present at the interface (Chattoraj & Birdi, 1984; Birdi, 2009). It is known that if a very small amount (a few micrograms per square meter of water surface) is applied to the surface, the surface tension changes markedly.

```
PURE WATER SURFACE.......
. . . . . . . . . . . . .WWWWWWWWWWWWWWWW. . . . . . . . . . . .
. . . . . . . . . . . . .WWWWWWWWWWWWWWWWW. . . . . . . . . . . .

WATER SURFACE WITH A LIPID........
. . . . . . . . . . . . .LLLLLLLLLLLLLLLLLLLLLLLLL. . . . . . . . . . . . .
. . . . . . . . . . . . .WWWWWWWWWWWWWWWWW. . . . . . . . . . . .
```

The lipid molecule moves under kinetic energy (**k T**). The change in surface tension is defined as *surface pressure*, Π.

Surface pressure

$$\Pi = \text{surface tension of pure water} -$$
$$\text{surface tension of water with lipid film} \qquad \text{(A.14)}$$

If the lipid is charged (which can be due to appropriate pH), then the monolayer exhibits interfacial charges, which can be studied by using the traditional **DLVO** theory. This monolayer method provides the most direct procedure of studying the state of such systems. Accordingly, the monolayer method has been a useful model system for complicated systems (such as emulsions, colloids, biological cell membranes). The different monolayer states that are extensively studied are described in the following sections.

Gaseous films

The lipid molecules are under the influence of different forces in the monolayer. The *gaseous* state is a film where only kinetic forces are present. This film would consist of molecules that are at a sufficient distance apart from each other such that lateral adhesion (van der Waals) forces are negligible. However, there is sufficient interaction between the polar group and the subphase that the film-forming molecules cannot be easily lost into the gas phase, and that the amphiphiles are almost insoluble in water (subphase).

The lipid film can be compressed by a suitable apparatus (Langmuir balance) (Figure A.7). It essentially consists of a Teflon bar which moves at the surface. When the area available for each molecule is many times larger than molecular dimension, the gaseous–type film [state 1] would be present.

As the area available per molecule is reduced (by moving the Teflon barrier, Figure A.8), the other states—for example, liquid–expanded [L_{ex}], liquid–condensed [Lco], and finally solid–like [S or solid–condensed]—would be present (Figure A.8).

The molecules will have an average kinetic energy, $1/2\ k_B T$, for each degree of freedom, where **k** is Boltzmann constant (= 1.372 10^{-16} ergs/T),

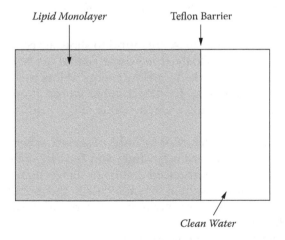

Figure A.7 A typical apparatus to study monolayer films (Teflon barrier separates the pure water surface from surface with lipid monolayer).

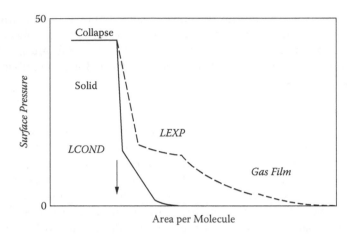

Figure A.8 The surface pressure versus area/molecule isotherms. The different states are gas region; liquid-expanded (LEXP) or liquid-condensed (LCOND); solid state; collapse state.

and T is the temperature. The surface pressure measured would thus be equal to the collisions between the amphiphiles and the float from the two degrees of freedom of the translational kinetic energy in the two dimensions. It is thus seen that the ideal gas film obeys the relation:

$$\Pi\,\mathbf{A} = \mathbf{k}_B\,T \text{ [ideal film]} \tag{A.15}$$

$$\Pi\,(\text{mN/m})\,A(\text{Å}^2 \text{ per molecule}) = 411\;(T = 298\;K)\;[\text{ideal film}] \tag{A.16}$$

Plots of $\Pi\,\mathbf{A}$ $(= \mathbf{k}\,T)$ versus $\mathbf{A}$ are found to show agreement with equation A.1. This is analogous to the three-dimensional gas law (i.e., $\mathbf{PV} = \mathbf{k}\,T$). At 25 C, the magnitude of $(\mathbf{k}_B\,T) = 411\;10^{16}$ ergs. If Π is in mN/m and $\mathbf{A}$ in Å², then the magnitude of $\mathbf{k}_B\,T = 411$. In other words, if one has a system with $\mathbf{A} = 400$ Å² per molecule, then the value of $\Pi = 1$ mN/m, for the ideal gas film.

In general *ideal gas* behavior is only observed when the distances between the amphiphiles are very large, and thus the value of Π is very small, < 0.1 mN/m. It is also noticed that from such sensitive data one can estimate the molecular weight of the molecule in the monolayer. This has been extensively reported for protein monolayers (Adamson & Gast, 1997; Birdi, 1989, 1999, 2009).

For example, insulin when added to surface of water at concentration of 0.58 mg m⁻², gave $\Pi = 0.1$ dyne cm⁻¹ (mN/m). These data can be analyzed as follows.

Monolayer data: insulin spread on water

$$\Pi\, \mathbf{A} = 0.58 \times 0.1 = 0.058 \text{ dyne cm}^{-1}\text{ m}^2\text{ mg}^{-1}$$
$$= 5.8\ 10^5 \text{ dyne cm--1 cm2 g--1}$$

From this, one can estimate the molecular weight of insulin.

Molecular weight = $\mathbf{R}\,\mathbf{T}/\Pi\,\mathbf{A}$

$$= (8.315)(10^7)(291)/(5.8\ 10^5)$$
$$= 42{,}000$$

This shows that insulin is present as a *hexamer*. The molecular weight of insulin monomer is 6000.

The latter observation requires an instrument with very high sensitivity, $\pm$ 0.001 mN/m. The Π versus A isotherms of n–tetradecanol, pentadecanol, pentadecyclic acid, and palmitic acid monolayers exhibit ideal behavior in the low Π region. The various forces that are known to stabilize the monolayers are mentioned as:

$$\Pi = \Pi_{kin} + \Pi_{vdw} + \Pi_{electro} \tag{A.17}$$

where Π_{kin} arises from kinetic forces, Π_{vdw} is related to the van der Waals forces acting between the alkyl chains (or groups), and $\Pi_{electro}$ is related to polar group interactions (polar group–water interaction, polar group–polar group repulsion, charge–charge repulsion).

When the magnitude of $\mathbf{A}$ is very large, then the distance between molecules is large. If there are no van der Waals or electrostatic interactions, then the film obeys the ideal equation. As the area per molecule is decreased, the other interactions become significant. The Π versus $\mathbf{A}$ isotherm can be used to estimate these different interaction forces. These analytical procedures have been extensively described in the current literature (Gaines, 1968; Adamson & Gast, 1997; Birdi, 1989, 1999, 2009).

The ideal equation has been modified to fit Π versus $\mathbf{A}$ data, in those films where co-area, $\mathbf{A}_o$, correction is needed (Birdi, 1989):

$$\Pi\,(\mathbf{A} - \mathbf{A}_o) = \mathbf{k}_B\,\mathbf{T} \tag{A.18}$$

In the case of straight-chain alcohols or fatty acids, Ao is almost 20 Å^2, which is the same as found from the X-ray diffraction data of the packing area per molecule of alkanes. This equation is thus valid when $\mathbf{A} \gg \mathbf{A}_o$. The magnitude of Π is 0.2 mN/m for $\mathbf{A}$ = 2000 Å^2 for ideal film. However, Π will be about 0.2 mN/m for $\mathbf{A}$ = 20 Å^2 for a solid-like film of a straight-chain alcohol.

Π versus (**A**) for a monolayer of valinomycin (a dodeca–cyclic peptide) shows that the relation as given in equation A.14 is valid (Figure 4.6). In this equation it is assumed that the amphiphiles are present as monomers. However, if any association takes place, then the measured values of (Π **A**) would be less than $k_B T < 411$, as has also been found (Birdi, 1989, 1999). The magnitude of $k_B T = 4 \ 10^{-21}$ J, at 25°C.

In the case of non-ideal films, one will find that the Π versus A data does not fit the relation in the equation. This deviation requires that one uses other modified equations of state. This procedure is also the same as one uses in the case of three-dimensional gas systems.

Liquid expanded and condensed films

The Π versus A data is found to provide a significant amount of detailed information about the state of monolayers at the liquid surface. In Figure A.8 some typical states are shown. The different states are very extensively analyzed and therefore will be described below.

In the case of simple amphiphiles (fatty acids, fatty alcohols, lecithins, etc.), in several cases, transition phenomena have been observed between the gaseous and the coherent states of films, which show a very striking resemblance to the condensation of vapors to liquids in the three-dimensional systems. The liquid films show various states in the case of some amphiphiles, as shown in Figure 4.6. (schematic). In fact, if the Π versus **A** data deviates from the ideal equation, then one may expect the following interactions in the film:

strong van der Waals
charge–charge repulsions
strong hydrogen bonding with subphase water

This means that such deviations thus allow one to estimate these interactions.

Liquid expanded films (L_{exp}). In general, there are two distinguishable types of liquid films. The first state is called the *liquid expanded* (L_{exp}) (Gaines, 1968; Chattoraj & Birdi, 1984; Adamson & Gast, 1997). If one extrapolates the Π–**A** isotherm to zero Π, the value of **A** obtained is much larger than that obtained for close-packed films. This shows that the distance between the molecules is much larger than one will find in the solid film, as will be discussed later. These films exhibit very characteristic elasticity, which will be described further below.

Liquid condensed films (L_{co})

As the area per molecule (or the distance between molecules) is further decreased, a transition to a *liquid condensed* (L_{co}) state is observed. These states have also been called *solid expanded* films (Adam, 1941; Gaines, 1968; Birdi, 1989, 1999; Adamson & Gast, 1997), which will be later discussed in further detail. The Π versus **A** isotherms of n–pentadecylic acid (amphiphile with a single alkyl chain) have been studied, as a function of temperature. Π–**A** isotherms for two chain alkyl groups, as lecithins, also showed a similar behavior.

Solid films

As the film is compressed, a transition to a solid film is observed, which collapses at higher surface pressure.

The Π versus **A** isotherms below the transition temperatures show the liquid to solid phase transition. These solid films have been also called *condensed films*. These films are observed in such systems where the molecules adhere to each other through the van der Waals forces, very strongly. The Π – **A** isotherm shows generally no change in Π at high A, while at a rather low **A** value a sharp increase in Π is observed. In the case of straight-chain molecules, like stearyl alcohol, the sudden increase in Π is found to take place at **A** $= 20 – 22\text{Å}^2$, at room temperature (that is much lower than the phase transition temperature, to be described later).

From these descriptions, it is thus seen that the films may under given experimental conditions show three first-order transition states: (1) transition from the gaseous film to the *liquid–expanded*, L_{ex}; (2) transition from the *liquid–expanded* (L_{ex}) to the *liquid–condensed* (L_{co}); and (3) transition from L_{ex} or L_{co} to the solid state, if the temperature is below the transition temperature. The temperature above which no expanded state is observed has been found to be related to the melting point of the lipid monolayer.

Collapse states

The measurements of Π versus **A** isotherms, when compressed, generally exhibit a sharp break in the isotherms, which has been connected to the *collapse* of the monolayer under the given experimental conditions. The monolayer of some lipids, such as cholesterol, is found to exhibit usual isotherm. The magnitude of Π increases very little as compression takes place. In fact, the collapse state or point is the most useful molecular information from such studies. It will be shown later that this is the only method that can provide information about the structure and orientation of amphiphile molecule at the surface of water.

However, a steep rise in Π is observed and a distinct break in the isotherm is found at the collapse. This occurs at $\Pi = 40$ mN/m and **A** $= 40$ Å^2.

This value of $\mathbf{A}_{co}$ corresponds to the cholesterol molecule oriented with the hydroxyl group pointing toward the water phase. Atomic force microscope (AFM) studies cholesterol as L–B films has shown that domain structures exist (see Chapter 5). This has been found for different collapse

Table A.4 Magnitudes of Ao for Different Film-Forming Molecules
on the Surface of Water

Compound	Ao ($Å^2$)
Straight-chain acid	20.5
Straight-chain acid (on dilute Hcl)	25.1
N–fatty alcohols	21.6
Cholesterol	40
Lecithins	ca. 50
Proteins	ca. $1m^2/mg$
Diverse synthetic polymers (poly–amino acids, etc.)	ca. $1m^2/mg$

Sources: Birdi, 1989; Gaines, 1968; Adamson & Gast, 1997.

lipid monolayers (Birdi, 2003). Different data have provided much information about the orientation of lipid on water, Table A.4.

It should be mentioned that monolayer studies are the only procedure that allows one to estimate the area per molecule of any molecule as situated at the water surface. In general the collapse pressure, Π_{col}, is the highest surface pressure to which a monolayer can be compressed without a detectable movement of the molecules in the films to form a new phase.

Interfacial changes of water due to monolayers
The presence of lipid (or similar kind of substance) monolayer at the surface of aqueous phase gives rise to many changes in the properties at the interface.

Surface potential (ΔV) of lipid monolayers
Any liquid surface, especially aqueous solutions, will exhibit asymmetric dipole or ions distribution at surface as compared with the bulk phase. If sodium dodecyl sulfate (SDS) is present in the bulk solution, then we will expect that the surface will be covered with SD⁻ ions. This would impart a negative surface charge (as is also found from experiments). It is thus seen that the addition of SDS to water not only changes (reduces) the surface tension but also imparts negative surface potential. Of course, the surface molecules of methane (in liquid state) obviously will exhibit symmetry in comparison with a water molecule. This characteristic can also be associated with the force field resulting from induced dipoles of the adsorbed molecules or spread lipid films (Adamson & Gast, 1997; Birdi, 1989).

The surface potential arises from the fact that the lipid molecule orients with polar part toward the aqueous phase. This gives rise to a change in dipole at the surface. There would thus be a change in surface potential when a monolayer is present, as compared with the clean surface. The surface potential, ΔV, is thus:

$$\Delta V = \text{surface potential}_{\text{monolayer}} - \text{surface potential}_{\text{clean surface}}$$
$$= V_{\text{monolayer}} - V_{\text{clean surface}} \tag{A.19}$$

The magnitude of ΔV is measured most conveniently by placing an air electrode (a radiation emitter: Po^{210} (alpha-emitter)) near the surface (ca mm in air) connected to a very high impedance electrometer. This is required since the resistance in air is very high, but it is appreciably reduced by the radiation electrode.

Since these monolayers are found to be very useful model biological cell membrane structures, it is thus seen that such studies can provide information on many systems where ions are carried actively through cell membranes (Chattoraj & Birdi, 1984; Birdi, 1989).

The transport of K^+ ions through cell membranes by antibiotics (valinomycin) has been a very important example. Addition of K^+ ions to the subphase of a valinomycin monolayer showed that the surface potential became positive. This clearly indicated the ion-specific binding of K^+ to valinomycin (Birdi, 1989).

Charged lipid monolayers

The spread monolayers have provided much useful information about the role of charges at interfaces. In the case of an aqueous solution consisting of fatty acid or sodium dodecyl sulfate (SDS), R–Na, and NaCl, for example, the Gibbs equation (2.40) may be written as (Adamson & Gast, 1997; Birdi, 1989; Chattoraj & Birdi, 1984):

$$-d\gamma = \Gamma_{RNa} \, d\mu_{RNa} + \Gamma_{NaCl} \, d\mu_{NaCl} \tag{A.20}$$

where γ is the surface tension, Γ_{RNa} and μ_{RNa}; Γ_{NaCl} and μ_{NaCl} are surface excess and chemical potentials of the respective species.

Further:

$$\mu_{RNa} = \mu_R + \mu_{Na} \tag{A.21}$$

$$\mu_{NaCl} = \mu_{Na} + \mu_{Cl} \tag{A.22}$$

and the following will be valid:

$$\Gamma_{NaCl} = \Gamma_{Cl} \tag{A.23}$$

and

$$\Gamma_{RNa} = \Gamma_R \tag{A.24}$$

Further, the following equation will be valid:

$$-d\gamma = \Gamma_{R\,Na}\, d\mu_{R\,Na} + \Gamma\, d\mu_{Na} + \Gamma\, d\mu_{Cl} \qquad (A.25)$$

This is the form of the Gibbs equation for an aqueous solution containing three different ionic species (e.g., R, Na, Cl). Thus, the more general form for solutions containing *i* number of ionic species would be:

$$-d\gamma = \Sigma\, \Gamma_i\, d\mu_i. \qquad (A.26)$$

In the case of charged film the interface will acquire *surface charge*. The surface charge may be positive or negative, depending upon the cationic or anionic nature of the lipid or polymer ions. This would lead to the corresponding *surface potential*, ψ, also having a positive or negative charge (Chattoraj & Birdi, 1984; Birdi, 1989). The interfacial phase must be electroneutral. This can only be possible if the inorganic counter-ions also are preferentially adsorbed in the interfacial phase.

The surface phase can be described by the Helmholtz double layer theory. If a negatively charged lipid molecule, R–Na+, is adsorbed at the interface, the latter will be negatively charged (air–water or oil–water). According to the Helmholtz model for double layer, Na+ on the interfacial phase will be arranged in a plane toward the aqueous phase. The distance between the two planes is of interest in such systems. The charge densities are equal in magnitude, but with opposite signs, Γ (charge per unit surface area), in the two planes. The [negative] charge density of the plane is related to the surface potential [negative], ψ_o, at the Helmholtz charged plane:

$$\psi_o = [4\,\pi\,\sigma\,\delta]/D \qquad (A.27)$$

where **D** is the dielectric constant of the medium (aqueous). According to Helmholtz double layer model, the potential ψ decreases sharply from its maximum value, ψ_o, to zero as δ becomes zero (Birdi, 1989, 1999). The Helmholtz model was found not to be able to give a satisfactory analysis of measured data. Later, another theory of the diffuse double layer was proposed by Gouy and Chapman: the interfacial region for a system with charged lipid, R–Na+, with NaCl.

As in the case of Helmholtz model, at a certain distance there will be an excess of negative charges, due to the adsorbed R– species. Due to this potential, the Na+ and Cl– ions will be distributed nonuniformly due to the electrostatic forces. The concentrations of the ions near the surface can be given by the Boltzmann distribution, at some distance x from this plane, as:

$$C^s_{Na+} = C_{Na+}\, [\exp -(\varepsilon\psi/kT)] \qquad (A.28)$$

$$C^s{}_{Cl^-} = C_{Cl^-} [\exp +(\varepsilon\psi/kT)] \tag{A.29}$$

where c_{Na^+} and c_{Cl^-} are the number of sodium and chloride ions per milliliter, respectively, in the bulk phase. The magnitude of ψ varies with x, from its maximum value, ψ_o. From the above equations we thus find that the quantities $c^s Na^+$ and $CsCl^-$ will decrease and increase, respectively, as the distance x increases from the interface, until their values become equal to c_{Na^+} and c_{Cl^-}, where ψ is zero.

The extended region of x between these two planes is termed as the *diffuse* or the **Gouy–Chapman double layer.**

The volume density of charge (per ml) at a position within this diffuse layer may be defined as equal to:

$$\varepsilon [cs+ - cs-] \tag{A.30}$$

which can be expressed by the Poisson relation:

$$d^2 \psi/d^2 x = -[4 \pi]/D \tag{A.31}$$

In this derivation it is assumed that the interface is flat, such that it is sufficient to consider only changes in ψ in the x direction normal to the surface plane. In the case of spherical particles, microsurfaces may also be treated as flat surfaces for such analyses.

The following relation can be derived from the above:

$$[d^2 \psi/d x^2] - d[d\psi/dx]/dx \tag{A.32}$$
$$= -[4 \pi N \varepsilon/1000 D] C$$

$$[e^{-\varepsilon/kT} e^{+\varepsilon\psi/kT}] \tag{A.33}$$

where c is the bulk concentration of the electrolyte.

In a circle of unit surface area on the charged plane, the negative charges acquired by the adsorbed organic ions (amphiphiles) within this area represent the surface charge density, σ:

$$\sigma = -\int \rho \, d x \tag{A.34}$$

$$= [D/4 \pi][d \psi/d x] \tag{A.35}$$

when integration is zero and infinity.

At x = 0, the magnitude of ψ reaches ψ_o:

$$\sigma = [2 D R T C/1000 \pi]^{1/2} [\sinh (\varepsilon \psi_o / 2 k T) \tag{A.36}$$

The average thickness of the double layer, $1/k$, (i.e. Debye–Hückel length), is given as (Chattoraj & Birdi, 1984):

$$1/k = [1000 \, D \, R \, T/8 \, \pi \, N^2 \, \varepsilon^2 \, c]^{0.5} \qquad (A.37)$$

At 25°C, for uni–univalent electrolytes, one gets:

$$k = 3.282 \, 10^{\,7} \, C^{-1/2} \, [\text{cm}] \qquad (A.38)$$

For small values of ψ, one gets the following relationships:

$$\sigma = [D \, R \, T \, k/2 \, \pi \, N \, \varepsilon] \, \sinh[\varepsilon \, \psi_o/2 \, k \, T] \qquad (A.39)$$

This relates the potential charge of a plane plate condenser to the thickness $1/k$. The expression based upon Gouy model is derived as:

$$\sigma = 0.3514 \, 10^5 \, \sinh \, [0.0194 \, \psi_o] \qquad (A.40)$$

$$= \Gamma \, z \, N \, \varepsilon \qquad (A.41)$$

where the magnitude of Γ can be experimentally determined, and the magnitude of ψ_o can be thus estimated. The free energy change due to the electrostatic work, Fe, involved in charging the double layer is as follows (Adamson & Gast, 1997; Chattoraj & Birdi, 1984; Birdi, 1989):

$$Fe = \int {}^{\psi o} \, \sigma \, d \, \psi \qquad (A.42)$$

By combining these equations one can write the expression for Π_{el} (Chattoraj & Birdi, 1984):

$$\Pi_{el} = 6.1 \, c^{1/2} \, [\cosh \, \sinh^{-1} \, (134/A_{el} \, C^{1/2})]^{-1}. \qquad (A.43)$$

The quantity $[k \, T]$ is approximately $4 \, 10^{-14}$ erg at ordinary room temperature (25°C), and $[k \, T/\varepsilon] = 25$ mV. The magnitude of Π_{el} can be estimated from monolayer studies at varying pH. At the iso-electric pH, the magnitude of Π_{el} will be zero (Birdi, 1989). These Π versus A isotherms data at varying pH subphase have been used to estimate Π_{el} in different monolayers.

Transport of ions in biological cell membranes

The most important biological cell membrane function is the transport of ions (such as Na, K, Li, Mg) through the hydrophobic lipid part of the BLM (Birdi, 1989, 1999). This property has relations to many diseases, such as infection and thus the activity of the antibiotics. These complicated biological processes have been studied by using monolayer model systems. For instance, *valinomycin* monolayers have been extensively investigated. The monolayers exhibited K-ion specificity, exactly as found in the biological cells. As is well known, cell membranes inhibit the free transport of ions (the alkyl chains of the lipids hinder such transport). However, molecules such as valinomycin assist in specific ion (K-ion) transport through binding (Gevod & Birdi, 1989; Chattoraj & Birdi, 1984).

Diverse constants

Avagadros number (N_A)	$6.0247 \ 10^{23}$
Velocity of light	$2.99793 \ 10^{10}$ cm sec^{-1}
Elementary charge (e_o)	$4.8029 \ 10^{-10}$ esu
	($1.60207 \ 10^{-19}$ coulomb)
Faraday (F)	96,520 coulomb/equivalent
Gas constant (R)	$8.3166 \ 10^7$ erg mol^{-1} deg^{-1}
Boltzmann's constant (k_B)	$1.3804 \ 10^{-16}$ erg deg^{-1}

1 cal = $4.186 \ 10^7$ erg = 4.1867 joule
$k \ T = 4.12 \ 10^{-21}$ J at 25C (298 K)
1 atm = $1.013 \ 10^5$ N m^{-2} Pa
$k \ T \ / \ e = 25.7$ mV at 25°C

References

Adam, N.K., *Chemistry of Surfaces*, Oxford University Press, 1941.

Adamson, A. and Gast, E., *Physical Chemistry of Surfaces*, Wiley, New York, 1997.

Birdi, K.S., *Lipid and Biopolymer Monolayers at Liquid Interfaces*, Plenum Press, New York, 1989.

Birdi, K.S., *Self-Assembly Monolayer Structures*, Plenum Press, New York, 1999.

Birdi, K.S., Ed., *Handbook of Surface and Colloid Chemistry*, 2nd Edition, CRC Press, Boca Raton, 2003.

Birdi, K.S., Ed., *Handbook of Surface and Colloid Chemistry*, 3rd Edition, CRC Press, Boca Raton, 2009.

Chattoraj, D.K. and Birdi, K.S., *Adsorption & the Gibbs Surface Excess*, Plenum Press, New York, 1984.

Gaines, G.L., *Insoluble Monolayers*, Wiley, New York, 1968.

Gevod, V. and Birdi., K.S., Melittin and 8-26 fragment monolayers, *Biophysical Journal*, 1079, 45, 1989.

Index

Milton Keynes UK
Ingram Content Group UK Ltd.
UKHW040053071024
449327UK00019B/533